SpringerBriefs in Geography

For further volumes:
http://www.springer.com/series/10050

David Higgitt
Editor

Perspectives on Environmental Management and Technology in Asian River Basins

 Springer

David Higgitt
Department of Geography
National University of Singapore
Singapore
e-mail: geodlh@nus.edu.sg

ISSN 2211-4165   e-ISSN 2211-4173
ISBN 978-94-007-2329-0   e-ISBN 978-94-007-2330-6
DOI 10.1007/978-94-007-2330-6
Springer Dordrecht Heidelberg London New York

Library of Congress Control Number: 2011937632

Springer is part of Springer Science+Business Media (www.springer.com)

Contents

Contents

Chapter 1
Environmental Management and Technology in Asian River Basins: Introduction

David Higgitt

Abstract Asian river basins are undergoing rapid transformation. Asian rivers are cradles of civilization, where interventions with natural river systems for irrigation, navigation improvement and flood control have a long history. It is not a coincidence that ancient civilizations arose in the lower reaches of large river basins, from the Nile to the Tigris–Euphrates, the Indus and the Yellow River. The ability to contain and utilize the waters and sediment-associated nutrients is a defining feature of ancient societies across many Asian river basins.

1.1 Environmental Challenges and Development in Asian River Basins

Asian river basins are undergoing rapid transformation. As Biswas and Tortajada (2011) have recently commented, water management in Asia will likely change more in the next 20 years than in the last 2000 years. Asian rivers are cradles of civilization, where interventions with natural river systems for irrigation, navigation improvement and flood control have a long history. It is not a coincidence that ancient civilizations arose in the lower reaches of large river basins, from the Nile to the Tigris–Euphrates, the Indus and the Yellow River. Karl Wittfogel's classic theory of hydraulic civilizations presented in *Oriental Despotism* (1957) charts the development of agricultural systems dependent upon, and subject to, large scale management of waterworks. The ability to contain and utilize the

D. Higgitt (✉)
Department of Geography, National University of Singapore,
1 Arts Link, Singapore 129792, Singapore
e-mail: geodlh@nus.edu.sg

D. Higgitt (ed.), *Perspectives on Environmental Management and Technology in Asian River Basins*, SpringerBriefs in Geography, DOI: 10.1007/978-94-007-2330-6_1, © The Author(s) 2012

waters and sediment-associated nutrients is a defining feature of ancient societies across many Asian river basins. Nevertheless, the present-day scale of land cover change and the extent and magnitude of contemporary water engineering is enormous in comparison to pre-industrial societies. The adaptations which ancient societies began to understand intrinsically in order to inhabit river valleys in monsoon regions, still remain as challenges to contemporary science and governance. And these challenges are amplified by critical environmental trends.

Key environmental trends include decreasing flows of water and sediment due to dam construction and increased water extraction, land use change (especially forest removal), encroachment and degradation of floodplain and wetland environments, increased water demand and concerns about water security, urban expansion and associated pollution of rivers and contamination of groundwater reserves. These challenges are being reframed by new approaches to water management which represent a transition from engineering-dominated approaches towards integrated water resource management (IWRM), from technology transfer to adaptive technologies. This book is concerned with exploring how some aspects of the reframing of approaches to water management may play out in the context of Asian river basins, where environmental protection and development goals are frequently seen in conflict. The Millennium Ecosystem Report (2005) recognizes that restoring and sustaining functional river ecosystems with high biodiversity is one of the greatest challenges facing society. On the other hand economic growth and the potential to transform rural livelihoods drive a perception that the water resources have been under-exploited. This situation is exemplified in the Mekong River Basin, where plans for large scale hydropower development have long been on the agendas of national governments, pushed forward by the World Bank, Asian Development Bank and a range of bilateral donors and hydropower companies. Middleton et al. (2009) describe the emerging 'electricity hunger' of riparian states heralded by a period of relative geopolitical stability and rapid economic growth. But the race to construct dams across a relatively undeveloped river basin has to be balanced by recognition of the river's existing value as a source of natural resources. In the case of the Mekong, the economic value of fisheries is significant but has often been underplayed in development plans (Middleton et al. 2009). Hydrological changes relating to dam construction and reservoir operation have significant negative impacts on fishery productivity and other riparian activities which, in turn, threaten food security.

Analysis of the driving forces contributing to contemporary environmental trends should therefore be accompanied by consideration of the coupled social trends which influence expectations of environmental management. Emerging expectations include greater accountability of organizations, widening participation in decision making, enhanced distribution of benefits among communities, sharing of risk and recognition of vulnerabilities (Lebel 2009). As such the adjoined environmental change and development challenges in Asian river basins require objectives of ecological sustainability and social justice to be balanced with further development of water resources. Some of the dilemmas arising can be illustrated by a brief consideration of flood management.

Evidence points to increasing incidence of floods globally but with particular significance in Asia. The construction of a Global Flood Inventory (Adhikari et al. 2010) cataloging the location, scale and impact of floods demonstrates the vulnerability of Asian cities. In a separate study of disaster impacts, 45% of water-related disaster fatalities and 90% of flood-displaced people in the period 1980–2006 were found to be in Asia (Adikari and Yoshitani 2009). The summer monsoon of 2010 inflicted unprecedented flood damage across wide swathes of Asia, most vividly in Pakistan and southern China. In response to flood hazard, a command and control strategy is commonly adopted where flood threat requires control through infrastructure and management from institutions. Reactions in the aftermath of flood incidents in Asian cities tend to focus on evaluating the per-formance of engineered systems. For example, a series of flash flood incidents in 2010 and 2011 which impacted the Orchard Road shopping area and Bukit Timah residential district of Singapore immediately triggered promises of engineering works to widen drain capacities and raise the level of affected roads. Through engineering ingenuity and foresight, the inherent problem of flooding in many urban areas has been controlled to a high degree. However, there are two areas of discord. The first is to consider how the modernist tendency towards a command and control approach is at odds with the traditional land use practices which adapted to monsoon flood regimes across Asia. Settlement patterns and cultural practices which evolved to accommodate annual flood cycles are disrupted by the pace of urbanization and environmental changes within hinterland basins. The second issue confronting the engineering-led approach is the moving target of flood management. The delivery of flood protection is compounded by several factors. These include the impact of rapid urbanization increasing runoff and flood peaks; the problem that the construction of flood protection encourages encroachment onto floodplains; and the concern that upstream interventions reallocates risk to other parts of the basin. The topic of how interventions may increase risk to vulnerable groups has received little attention (Lebel and Sinh 2007). Similarly, strong migration from rural areas to cities resulting in expansion of slums into flood risk areas is a neglected topic (Adikari et al. 2010). Two further aspects which make flood management a moving target are limited knowledge about the range of extreme events and the potential impacts of climate change. Measurement of water discharge has a limited history in most Asian countries though documentary records of flood events are impressive in China and Japan. Ultimately, appraisal of the probability of events exceeding a certain threshold is based on limited data. The ability to derive longer-term information about past floods through interpretation of past sedimentary deposits is one possibility to extend knowledge. Though the past incidence of floods and droughts has been a feature of life in Asia, there is growing evidence that the range of extremes and frequency of hazard events has been increasing because of climate change. The extent to which a command and control management style can keep up with climate change is also called into question. Finally, as mentioned above, changing expectations from societies to sustain functional river systems challenge the reliance on infrastructure.

Juxtaposed with the increasing concern about flood hazard across several Asian cities, there are signs of an emerging international movement towards integrated river basin management. In several Western countries, most notably in USA and Australia, societal demands for river rehabilitation—the improvement and maintenance of functional river ecosystems—have resulted in large-scale initiatives to recover or restore damaged aquatic ecosystems. In the European Union, the Water Framework Directive provides a blueprint for maintaining water quality, protecting high value ecosystems and recognizing the ecosystem services provided by rivers. These initiatives demand a new approach to river basin management, where interdisciplinary activity across the water-related earth sciences, particularly involving engineers, hydrologists, geomorphologists and ecologists, is tied more effectively to social science perspectives, and these understandings are appropriately framed within an institutional context. There are signs that this new paradigm is establishing itself in Asia. The Asian River Restoration Network (ARRN) is a non-governmental organization, established in 2006 to promote the exchange of knowledge and technology, particularly in relation to the Asian monsoon. National networks have been established in China, Korea and Japan, the latter providing secretariat support for the Asian network. ARRN argues for the necessity of countries in the Asian monsoon region to develop and share strategies for river restoration given the dense populations, regime of frequent flooding and abundant rice paddy.

1.2 Integrated Water Resource Management

The accelerated transformation of the physical environment coupled with changing expectations for environmental management (outlined above) lays the template for a multitude of regional, national and local challenges facing the water sector in Asia. Many communities still lack access to a safe source of water and reliable sanitation systems. Climate change exerts further challenges on the ability of infrastructure to cope with flood risk, threatens water scarcity in some areas due to reduced precipitation or groundwater recharge and accentuates risk of contamination of water resources. Water scarcity is a major concern though some commentators (e.g. Biswas 2009) express cautious optimism that present knowledge, experience and technology will be sufficient to address problems if inefficient management practices can be improved. In an extended analysis of the Asian Development Bank's Asian Water Development Outlook, Biswas and Seetharam (2008) argue that by concentrating on the tricky issues of capacity building, attaining political will and appropriate investment the problems of overcoming poor management of water resources in some parts of Asia can be solved. Thus it is not physical scarcity but poor water management which lies at the heart of the problem. This requires innovative approaches and new mindsets.

IWRM represents an attempt to frame water management in such a new mindset to focus on human well being and environmental sustainability. Snellen and Schrevel (2004) have traced the history of United Nations involvement in IWRM back to the late 1950s, taking its precedent from the Tennessee Valley Authority established in the USA in 1933. Here is explicit recognition that engineering design alone was not sufficient to bring about desired improvements in livelihoods but needed to be accompanied by consideration of other resource use, such as access to finance, transport and fertilizer in irrigation schemes. From a secondary concern to link water planning with other sectoral interests, IWRM has been propelled forward in the last two decades as a means to promote coordinated development and management of water, land and other resources addressing the goals of efficient and economic use of water, equity in allocation and access across different social groups and environmental sustainability to protect and improve the ecological value of the environment. Arising from the 1992 International Conference on Water and the Environment, the so-called Dublin Principles emphasize that water is a finite and vulnerable resource vital in sustaining life, the need to base management on a participatory approach involving users, planners and policy makers at all levels, the important but often neglected role played by women in managing and safeguarding water and the recognition that water should be treated as an economic good. This led to the establishment of the Global Water Partnership (GWP) in 1996 which has adapted and elaborated these principles, but also recognized that nation states at different stages of development will need to adapt IWRM principles to local contexts.

1.3 Science and Technology Studies: A Framework

If the understanding and management of water resources has traditionally been regarded as the realm of scientists and engineers, the emerging debates about alternative approaches to river basin management call for greater interdisciplinary engagement. Science and Technology Studies (or Science, Technology and Society; STS) is an established interdisciplinary field which recognizes the need for social scientists to investigate the huge influence of science and technology permeating every aspect of political, social and cultural life. Emerging dilemmas confronting societies such as climate change or environmental sustainability require engagement with science and technology, yet mix politics, science and popular knowledge. Many aspects of water resource management involve concepts of knowledge transfer where a piece of technology or an example of best practice is introduced which can be 'learnt' and applied to new local situations. Science and Technology scholars (e.g. Latour 1987) argue that technology is socially constructed and is made and remade as the dissemination of the technology evolves into new and unplanned applications. Pahl-Wostl et al. (2011) notes that the development and application of IWRM has mirrored emerging theories of knowledge and management, such as notions of extended peer communities and socially robust science. Thus STS provides a potential framework for examining how scientific research on hydrological and ecological dynamics of rivers can be

contributed, engaged and transformed. Recent interest in adaptive management
and social learning provide an illustration.

Concepts of adaptive management arose from interest in non-linear change and
complexity in ecological systems. It can be defined as an iterative process for
improving decision making, policy and practice through learning from outcomes
of previously implemented strategies. Tackling the enormous challenges of sustainability and cross-sectoral planning is more appropriately approached by
regarding the mix of human, technology and environmental components as part of
a complex adaptive system where uncertainty is inevitable (Pahl-Wostl 2007).
In juxtaposing elements of traditional command and control strategies against
elements of IWRM (Zevenbergen et al. 2008) several conceptual features of
adaptive management can be identified. The tendency to focus on reducing
uncertainties by devising technical solutions to narrowly defined problems is a
traditional characteristic of water resource management now challenged by the
notion that co-evolving complex adaptive systems are irreducible and require an
approach that embraces uncertainty as a key feature. Hence the traditional strategies to control change to preserve a status quo or focus on reducing a specific risk
can be challenged by strategies to enhance ability to adapt to uncertainty and a
focus on reducing vulnerability—concepts of adaptiveness and resilience. A linear
approach to sequential planning processes is challenged by a willingness to realign
content and process according to context—recognition of complexity; by an
extension of the typical timeframe for planning towards longer term horizons
framed by concepts of sustainability. The tradition of top-down management by
strong institutions is challenged by the encouragement of bottom-up initiatives
focused by strategic leadership. Participation is a fundamental ingredient.

Despite the growing intellectual debate on adaptive management there is
arguably limited evidence of genuine shifts in practice (Gregory et al. 2011). The
governance frameworks underpinning management practice impose inertia to limit
transition. Thus there is increasing attention to ideas of social learning arising from
literature in sustainability science and organizational management research. Social
learning and adaptive management are linked by the need to develop capacity of
relevant authorities (e.g. planners, policy makers, experts, interest groups) and other
stakeholders to negotiate goals and translate these into actions. Learning is
achieved within the context of the network of actors. Adaptive management
requires a human-technology-environment system (i.e. a network of actors) to adapt
in response to, or in anticipation of changes in the environment (Lebel et al. 2010).

In Australia, where empathy towards new paradigms of river repair and
adaptive management has been generally established, the transition to new
governance frameworks remains challenging. In an empirically based study,
Farrelly and Brown (2011) examined a series of local scale experiments in
Queensland, Victoria and Western Australia. They argue that in order to accommodate uncertainty, organizational cultures must be willing to embrace experimentation as a precursor to learning. In practice the rigid regulatory frameworks
within which centralized management systems operate provide an intimidating
regime for practioners to experiment with novel innovations or technologies.

A fear of failure or risk aversion is prevalent as this has implications for financial considerations, reputation, liabilities and loss of future opportunities (Farrelly and Brown 2011). Building on the concept of triple-loop learning, the single loop or technical learning can be regarded as a process of selecting actions within an existing set of assumptions; the double-loop or conceptual learning revisits assumptions and perceptions of the defined problem; the triple-loop or social learning reconsiders underlying values and beliefs providing the potential for transformation (Pahl-Wostl et al. 2011). Applied to Asia, the openness of organizational structures to experimentation and innovative approaches, the capacity to build stakeholder networks and the profile of emerging paradigms of river repair and ecological sustainability as central pillars of water management may be rather different from experiences in Australia, North America and Europe. There are signs that many of these ideas are being picked up, debated, shared and transformed in Asian contexts. But empirical analysis of the effectiveness of learning processes and the diffusion of innovation in Asian river basins has been quite limited. Questions about how the interaction of science, technology, policy and governance plays out to address the multiple challenges of water resource management, needs further attention. STS may provide theoretical insights for framing future debate.

1.4 Themes and Case Studies

The volume comprises an introduction and five chapters. In Chap. 2, Gary Brierley and Carola Callum reframe approaches to river repair within emerging theories of in ecology and earth science which regard nature as a complex adaptive system replete with inherent uncertainties. They trace the development of ideas about complex adaptive systems over the past 50 years which has some resonance with the gradual evolution of IWRM principles among international agencies. These ideas are mirrored in the four case study chapters which deal with specific examples of governance, science or policy issues in Asian river basins. The studies descend from the alpine meadows of Qinghai province China, the source of three great Asian rivers—the Yangtze, Yellow and Mekong, to consider governance issues in the lower Mekong, adaptation of water technologies by local communities in western India and analysis of the deforestation-erosion debate in the uplands of Java, Indonesia. Naturally, these case studies provide just a glimpse of the many problems and challenges confronting Asian river basins but they do cut across key themes of improving scientific understanding of hydrological systems, appreciating the making and remaking of technologies transferred from western to Asian contexts, the emergence of joined-up policy interventions to link nature conservation within a framework of water management.

Chapter 3 visits the Mekong Basin. Philip Hirsch examines the shift from 'hardware' to 'software'-driven approaches. Hardware refers to the engineering-based approach of the command and control school with its attendant focus on infrastructure. The software dimension is the focus on governance and IWRM. Hirsch considers the implications of new governance approaches for the implementation of large scale infrastructure plans. The Mekong holds a special place in discussions of IWRM in Asia because of its transboundary status and the comparatively long history of the Mekong River Commission (MRC). In a recent analysis of social learning, Lebel et al. (2010) charts the development of stakeholder networks and engagement and the progress of the MRC in learning how to 'do' public participation. Hirsch illustrates his chapter with three cases examining bilateral relations between Vietnam and Cambodia on the Sean tributary, the much-debated issue of upstream Chinese dams and the sustainability-driven agenda of the Thai water grid.

In Chap. 4, Angela Barbanente, Dino Borri and Laura Grassini adopt and approach informed by STS to examine the hybrid adoption of water resource technologies in India. The first example is from agricultural communities in the arid areas of Gujarat, where local adaptation of a technology to recharge groundwater triggered hydrogeological investigation. The second example examines a slum upgrading project in Ahmedabad where the improvement of water supply facilities resulted in a range of additional and innovative adaptations by the local communities. Thus the intended outcomes of technology transfer spin off into unanticipated innovations as local communities combine indigenous and external technologies in hybrid forms.

Moving from India to Indonesia, Chap. 5 by Anton Rijsdijk considers the problem of determining sediment dynamics in an upland catchment in Java. Ostensibly a geomorphological research project to evaluate sediment yield in relation to land use, the results throw light onto a controversial but often over-simplified problem encountered in Indonesia and other steepland environments in Southeast Asia. This is the blame game where flood disasters are attributed to land use change (particularly to illegal logging) in upland areas. The logic of associating deforestation with downstream increases in flood peaks is immediately apparent but it is surprising how few studies have conclusively demonstrated the magnitude of hydrological changes. As Meigh and Barlett (2010) have demonstrated from consultancy projects in Indonesia and the Philippines, there are multiple factors contributing to flooding in addition to poor land use upstream. Floodplain encroachment and garbage accumulation and other restrictions in urban channels are problematic. Rijsdijk uses a long-term paired catchment approach to compute sediment yields in the upper (forested) and lower (agricultural) parts of the two catchments. The data indicate that sediment yields is significantly enhanced in areas where the forest is removed but also that official data are likely to have underestimated sediment yields because of inadequate sampling procedures.

The final chapter (Chap. 6) by Xi-Lai Li et al. goes to the roof of the world. The Sanjiangyuan (Three Source Zone) Natural Reserve in Qinghai Province nurtures three great rivers: the Yellow, Yantgtze and Mekong. Climate change, increased human activity and overgrazing have degraded significant areas of natural grasslands in Qinghai which in turn reduces the carrying capacity and threatens the livelihoods of local communities dependent of herding. This degradation has been accompanied by reductions in discharge in the headwaters, at least in part due to retreating glaciers. Under a national policy to increase the economic potential and well being of inhabitants in its western provinces, the protection of biodiversity and the restoration of degraded land are set forth as important objectives. Li et al. describe the ecological conservation imperatives and their connection to integrated river basin management.

The collection of papers began life in a session organized for the NUS Centennial Conference on "Asian Horizons: Cities, States and Societies". It was a deliberate attempt to bring together academics concerned with rivers and water resources from either side of the science–social science divide in an attempt to examine shared interests, concerns and obstacles to engagement. The obstacles turned out to be mainly logistical as the collected papers destined for a journal special issue languished with more than one unresponsive editor. I am grateful to Petra van Steenbergen for rescuing the project and facilitating its presentation in the SpringerBriefs series in Geography. Each of the contributed chapters has been through a full peer review and revision process. In this endeavour, I acknowledge the contributions of Radhan D'Souza (Waikato University, New Zealand); Feng Yiming (Chinese Academy of Forestry, China); Carl Grundy-Warr (National University of Singapore); Stuart Harris (University of Calgary, Canada); Louis Lebel (Chiangmai University, Thailand); Emma Mawdsley (University of Cambridge, UK); John Rowan (University of Dundee, UK); Michael Stewardson (Australian National University, Australia); Ross Sutherland (University of Hawai'i, USA); and Bob Wasson (Charles Darwin University, Australia). Most of all I am grateful to the authors whose patience has been severely tested through this process.

References

Adhikari P, Hong Y, Douglas KR, Kirschbaum DB, Gourley J, Adler R, Brakenridge GR (2010) A digitized global flood inventory (1998–2008): compilation and preliminary results. Nat Hazards 55:405–422

Adikari Y, Yoshitani J (2009) Global trends in water-related disasters: an insight for policymakers. United Nations World Water Development Report 3, Water in a Changing World. UNESCO, Paris

Adikari Y, Osti R, Noro T (2010) Flood-related disaster vulnerability: an impending crisis of megacities in Asia. J Flood Risk Manag 3(3):185–191

Biswas AK (2009) Water management: some personal reflections. Water Int 34(4):402–408

Biswas AK, Seetharam KE (2008) Achieving water security for Asia. Int J Water Resour Dev 24(1):145–176

Biswas AK, Tortajada C (2011) Water quality management: an introductory framework. Int J Water Resour Dev 27(1):5–11

Farrelly M, Brown R (2011) Rethinking urban water management: experimentation as a way forward? Glob Environ Chang. doi:10.1016/j.gloenvcha.2011.01.007

Gregory C, Brierley G, Le Heron R (2011) Governance spaces for sustainable river management. Geogr Compass 5(4):182–199

Latour B (1987) Science in action: how to follow scientists and engineers through society. Open University Press, Milton Keynes

Lebel L (2009) The global environmental change and development nexus in Southeast Asia. In: Lebel L, Snidvongs A, Chen AC-T, Daniel R (eds) Critical states: environmental challenges to development in monsoon Southeast Asia. Strategic Information and Research Development Centre, Petaling Jaya, Malaysia, pp 3–17

Lebel L, Sinh BT (2007) Politics of floods and disasters. In: Lebel L, Dore J, Daniel R, Koma YS (eds) Democratizing water governance in the Mekong region. USER, Chiang Mai University, Chiang Mai, pp 37–54

Lebel L, Grothmann T, Siebenhüner B (2010) The role of social learning in adaptiveness: insights from water management. Int Environ Agreem 10(4):333–353

Meigh JD, Barlett JW (2010) Integrated river basin management in Southeast Asia. J Flood Risk Manag 3(3):201–209

Middleton C, Garcia J, Foran T (2009) Old and new hydropower players in the Mekong Region: agendas and strategies. In: Molle F, Foran T, Kakonen M (eds) Contested waterscapes in the Mekong Region. Earthscan, London, pp 23–54

Pahl-Wostl C (2007) Transition towards adaptive management of water facing climate and global change. Water Resour Manag 21(1):49–62

Pahl-Wostl C, Jeffrey P, Sendzimir J (2011) Adaptive and integrated management of water resources. In: Grafton RQ, Hussey K (eds) Water resources planning and management. Cambridge University Press, Cambridge, pp 292–310

Snellen WB, Schrevel R (2004) IWRM: for sustainable use of water. Fifty years of international experience with the concept of integrated water resource management. Background document to the FAO/Netherlands Conference on Water for Food and Agriculture. Ministry of Agriculture, Nature and Food Quality, Amsterdam

Wittfogel KA (1957) Oriental despotism: a comparative study of total power. Yale University Press, New Haven

Zevenbergen C, Veerbeek W, Gersonius B, Van Herk S (2008) Challenges in urban flood management: travelling across spatial and temporal scales. J Flood Risk Manag 1(2):81–88

Chapter 2
Environmental Science and Management in a Changing World

Gary Brierley and Carola Cullum

Abstract Recognition of the pervasive impact of human activities on the natural world has prompted a reframing of approaches to environmental repair. Disentangling threads in the evolution of theoretical environmental science and management practice over the last 50 years enables practitioners to position contemporary programmes within a coherent conceptual framework. Over the last 50 years, environmental management strategies have moved beyond concern solely for utilitarian values to recognizing the importance of biodiversity management and associated ecosystem services. Initially, these programmes focussed on management of single issues, emphasizing concerns for endangered species and conservation in designated reserves that retained notionally pristine areas. The transition to an ecosystem approach to environmental management marked the recognition that there are too many species to attempt to save them one at a time and that conservation efforts must focus upon whole ecosystems. However, the ecosystem approach retained a vision of nature in equilibrium and management initiatives were framed in terms of environmental reference conditions and adjustments around a mean condition. Emerging theories in both ecology and earth sciences view nature as a complex adaptive system, often involving non-linear relationships and stochastic events which lead to outcomes that are unpredictable in time and space. Explicit recognition of inherent uncertainties in the character and behaviour of environmental systems has prompted a shift towards application of adaptive and participatory management principles. Recognizing directly the imperative to integrate scientific thinking with social-ecological

G. Brierley (✉)
School of Environment, University of Auckland, Private Bag 92019,
Auckland, New Zealand
e-mail: g.brierley@auckland.ac.nz

C. Cullum
Centre for Water in the Environment, University of Witswatersrand,
Johannesburg, South Africa

D. Higgitt (ed.), *Perspectives on Environmental Management and Technology in Asian River Basins*, SpringerBriefs in Geography, DOI: 10.1007/978-94-007-2330-6_2, © The Author(s) 2012

considerations, post-normal scientific practice embraces a more inclusive approach to the management of complex adaptive systems, striving to maximise the resilience of any given system. Prospective opportunities for geographers in meeting the needs of this emerging framework are considered.

Keywords Environmental management · Ecosystem management · Biodiversity · Post-normal science

2.1 Introduction

The last 50 years have seen a dramatic shift of perspective in our scientific understanding of the natural world. The traditional view of the 'balance of nature' was embedded in conceptual frameworks that built upon equilibrium notions and related conceptualisations of succession, climax communities and notional 'endpoints' of ecosystem trajectories (Botkin 1989; Wu and Loucks 1995). Over the last decade or so, this view has been supplanted by a vision of nature as a complex adaptive system, characterised by non-linear relationships, random events and interactions that generate uncertainties and discontinuities (O'Neill 2001; Perry 2002). It is now recognised that environmental interactions and responses are place and time specific, so that different catchments will react in different ways to similar interventions, depending on their climatic and geological context, spatial configuration and their history of evolution and land use (Brierley and Fryirs 2005).

As theoretical perspectives have evolved, so too have their applications in environmental management. For example, as long as natural systems were understood to be generally stable if unaltered by humankind, then conservation efforts could aim to preserve or restore a pristine state. Conceptualizations of nature in terms of mechanical models and linear, causal relationships enabled indicators to be used to evaluate environmental health and to track progress towards restoration goals. However, a new, emerging paradigm views the natural world as a complex system that is not necessarily in equilibrium. Multiple states are possible within the same boundary conditions, and the same impact may have very different consequences in different contexts. Such thinking undermines the use of pristine conditions as reference points and broadly applied indicators. In response, adaptive and participatory management frameworks provide the most reliable platform to work with, and plan for, uncertainties (Walters and Holling 1990). They recognise explicitly that our understanding will never be complete, and that sustainable environmental management is inherently dependent upon societal engagement and empowerment in the design, implementation and maintenance of management activities (Rhoads et al. 1999).

The shift in theoretical perspective not only has far-reaching implications for its practical application in management, but also for the way environmental science needs to be conducted in order to usefully inform policy and decision making.

As noted by Ravetz (1999, p 649), in such 'issue-driven' science "typically facts are uncertain, values in dispute, stakes high, and decisions urgent.... the previous distinction between 'hard', objective scientific facts and 'soft', subjective value-judgements is now inverted. All too often, we must make hard policy decisions where our only scientific inputs are irremediably soft". Funtowicz and Ravetz (1993) argue that 'normal' science, based on reductionist theories supported by experimental evidence and notionally value-free deductive reasoning, needs to be replaced by 'post normal science', characterised by uncertainty analysis and management, lay participation and the integration of scientific knowledge not only across disciplines, but also across the lay-expert divide.

The new scientific paradigm of natural systems thus presents significant challenges to scientists and managers alike. Scientists cannot provide secure predictions to inform managerial policy and interventions, solutions are not universally applicable, desired outcomes are negotiable and the involvement of local communities and stakeholders is essential. In many instances, institutional structures and practices have only recently been reformed in light of international, national and local environmental legislation developed during the 1990s such as the 1992 Rio Convention on Biological Diversity (United Nations Environmental Program 1992).

Although there have been many statements of the principles of adaptive management and post normal science (e.g. Kay et al. 1999), uptake has been slow in practice as practitioners are often trapped within the bounds of discipline-bound academic institutions and legislatively-bound environmental management agencies. Inevitably, there is a lag between the evolution of scientific thinking and management practice. The nature and extent of this dislocation varies markedly from place to place, as institutions themselves respond to their own histories and associated cultures (framed in light of prevailing policy imperatives). As in natural systems, the drivers of change can lead to different responses in different institutional situations. Context and history are important constraints on potential outcomes. In this changing world, variability in both natural systems and in the people that use and inhabit them, present significant challenges to environmental scientists and managers.

In seeking to unravel some of these complexities, it is important to ground the evolution of conceptual developments and managerial responses, thereby providing the context with which to position the status and prospects for practitioners and institutions that work in environmental management. This manuscript endeavours to disentangle the strands of scientific and managerial practice in environmental management in the western world over the last 50 years, drawing especially upon references from the biodiversity and river management arenas. Three phases in the development of conservation theory and practice are identified. 'Command and control' environmental management and early conservation efforts focussed around single issues and was informed by ecological theory that stressed the composition of ecosystems. This gave way to a more holistic 'ecosystem' approach, in which functional elements were emphasized. The contemporary view focuses on system dynamics and complexity. Each paradigm implies quite different relationships between scientists, managers, society and the natural world, reflected in distinct styles of conservation practice and environmental management.

It is hoped that this classification of managerial approaches and their theoretical foundations will help practitioners to critically appraise current and future initiatives. Post normal science and adaptive management techniques are overviewed as responses to the uncertainty implicit in complex, adaptive ecosystems. In the concluding section, we suggest that geographers are well placed to take a leading role in the uptake of these emerging approaches to environmental management.

2.2 Interplay Between Ecological Theory and Management Over the Last 50 Years

Over the last half-century, environmental management practices have moved beyond approaches that treat components of natural systems in isolation towards more holistic approaches that recognise ecosystem connectivity and the need to encompass both natural processes and those resulting from human intervention. Three phases can be differentiated over this era, each characterised by a different understanding of the concept of biodiversity (Table 2.1). A clear appreciation of these differences is critical to constructing a coherent conceptual framework that does not conflate conflicting views of theory and practice.

2.2.1 Single Issue Focus in the era of 'Command and Control'

The notion of 'biodiversity' emerged in the 1960s as a theme linking scientific study of biological heterogeneity, popular concern with increasingly visible environmental degradation and the establishment of professional environmental managers and conservationists (Gaston 1996). Initially, the focus was on ecosystem composition, with scant attention to spatial or functional context. Natural systems were generally viewed as static, or in dynamic equilibrium once succession had reached the 'climax' stage (Wu and Loucks 1995; O'Neill 2001; Perry 2002, inter alia). Natural phenomena such as flood, fire and earthquakes were seen as 'disturbances' that turned back the clock, resetting the system to an earlier successional stage, rather than as integral components of the system. Human intervention was seen in a similar way, such that the aim of conservation efforts was to preserve pristine landscapes or to restore them to a 'natural' state, which it was assumed would then persist indefinitely (Sprugel 1991). It was believed that scientific progress would ultimately reveal the mechanisms of natural systems, such that remedial engineering could reverse adverse human impacts.

In this era, species were considered as the 'units of biodiversity' (Claridge et al. 1997). A plethora of quantitative biodiversity indices was developed, designed either to evaluate the relative conservation value of different areas or to compare measurements at different times when monitoring the impact of land use changes or restoration programmes. Most centred on measurements of species richness and

Table 2.1 Developments in ecological theory and their expression in environmental management

	Single issue	Ecosystem	Complex adaptive system
Biodiversity: concepts and emphasis	Collection of individual species	Interactions between species and physical habitat—the field of dreams hypothesis	Complexity—non linear relationships, stochasticity, historical path dependence, contingency, unpredictability, uncertainty, alternative states possible
	Structure-composition	Function	Dynamics-process
Human relationship to nature	Control and exploitation of resources Management of risk	Management of 'natural' systems	Humans viewed as part of social-ecological systems
Management goals	Command and control; Make nature conform to human needs; Uniform, generally applicable solutions; Conservation of individual species	Pressure-State-Response Conservation of entire (local) systems; Reference state—pristine condition (end points)	Adaptive management; Place-specific goals negotiated between stakeholders; System sustainability, maintenance of ecosystem services
Theoretical framework	Stability	Dynamic equilibrium, disrupted by disturbance	Complexity, self-organisation; Natural variability, disturbance part of the system
	Individual components	Web of interactions centred around species/area of interest	Nested hierarchies, level characterised by relative temporal/spatial scales
Dominant disciplines	Engineering	Ecology (with links to other natural sciences)	Transdisciplinary—human and natural sciences
Methods	Single- purpose engineering projects	Indicators of system health, Targets enshrined in protocols/legislation	Thresholds of probable concern; Targets locally negotiated
	"Normal" science	Cross-disciplinary, summative approach	"Post-normal" science. Integrative, transdisciplinary, systems-based thinking about social-ecological interactions
	Mechanistic models	Process based models with predictive ability to forecast outcome	Simulation models suggesting possible outcome of processes; Spatially explicit models and GIS; Foresighting of scenarios
	Biodiversity indices	Keystone/umbrella species	System resilience

evenness, although some indices also incorporated measures of rarity or endemism (Ricotta 2005). Thus the assessment of biodiversity involved counting the number of species present in a particular area and their relative abundance. To circumvent the need to identify and count the vast number of species present in even a very small area, surrogate taxa were sought as indicators of the total level of diversity likely to be present (e.g. Pearson 1994). However, several studies showed that no single component could consistently indicate overall biodiversity, since the diversity of different groups can vary independently at almost every biological scale, from genes through to whole ecosystems (e.g. Prendergast 1997; Heino et al. 2005). Biodiversity indices are now also criticised as gross oversimplifications, failing to distinguish the relative functional or conservation importance of different species (Noss 1990; Hooper et al. 2005; Ricotta 2005).

Given the challenge of measuring biodiversity, it is no surprise that early conservation efforts focussed on the preservation of individual threatened species (Simberloff 1998). Underpinned by emerging theories of population dynamics, management initiatives concentrated on large, relatively easily sampled, well-studied and charismatic organisms. For example, in 1991, 51.5% of all US government spending on the protection of endangered species was concentrated on only seven taxa (Losos 1993; Simberloff 1998).

Single-issue thinking also characterised the 'command and control' era of river management that prevailed in the wake of the industrial revolution (Cosgrove and Petts 1990; Holling and Meffe 1996; Hillman and Brierley 2005). Driven by political and social concerns for protection against natural hazards and the development of resources, this engineering-based approach was framed in terms of deterministic thinking about river stability and natural processes. Grounded in the reductionist scientific method derived from physics and chemistry, it assumed that nature can be described in terms of linear causal relationships and mechanistic models. The 'command and control' approach also positioned rivers outside their landscape context, treating individual components of the system in isolation, such that ecosystems were often harmed as a result of unanticipated knock-on effects (Richter and Postel 2004). Strategies applied over relatively short reaches of river targeted stability, conformity and simplicity, epitomised by the construction of uniform, homogenous, hydraulically smooth channels, dams, drained wetlands and fixed flood barriers. The range of natural variability was limited as the system was forced to conform to human needs, often resulting in a decrease in system resilience to disturbance and heightening the effect of extreme events (Holling and Meffe 1996). Many river conservation projects also failed as symptoms were treated without addressing the root cause (e.g. Hilderbrand et al. 2005).

2.2.2 A Holistic Ecosystem Approach

By the 1990s, the conservation focus on individual species was under attack as extinction rates continued to soar. It was recognised that there are just too many species to attempt to save them one at a time and that conservation efforts must be directed at whole ecosystems. A new view of biodiversity was emerging, in which the composition, structure (spatial pattern) and function (ecological and evolutionary processes) of an ecosystem both determined and constituted its biological diversity (Noss 1990; Franklin 1993). Multiple biotic and spatial scales were considered, from genetic diversity through to the variability of whole ecosystems (Wegner et al. 2005). Whilst acknowledging the value of a species approach as a valid 'emergency-room' tactic to bring species back from the verge of extinction, environmental management efforts started to focus on the preservation or restoration of entire habitats and their ecological 'integrity' rather than on individual species or the overall number of species present. Such an approach served to protect the numerous, small and often undescribed species of invertebrates, fungi and bacteria that carry out critical ecosystem functions such as decomposition or nitrogen fixation (Franklin 1993). In contrast to basic biodiversity indices, analyses framed in terms of the ecosystem approach recognise that unmeasured external factors such as invasion by weedy or exotic species may compromise the ecological functioning and evolutionary potential of an ecosystem (Noss 1990).

Methods of biodiversity assessment changed to reflect the new theoretical insights. Rather than attempting to enumerate all species present in an area, or to identify taxa or species whose presence would signal diversity in all other groups of organisms, indicator taxa or species were sought that could detect and monitor compositional, structural and functional biodiversity at multiple levels of organisation (Noss 1990). Concepts such as keystone species (Paine 1995), umbrella species (Launer and Murphy 1994; Lambeck 1997), functional guilds (Severinghaus 1981; Block et al. 1987) and ecosystem engineers (Jones et al. 1994) were developed by ecological theorists and used by conservation managers to justify the selection and establishment of reserves.

Many environmental managers, policy makers and members of the public welcomed the advent of ecological indicators that enabled complex systems to be summarised in ways which could be simply communicated. Indicators could be used to signal progress towards targets based on notions of a pristine system or high-quality reference sites (Niemi and McDonald 2004). A Pressure-State-Response framework (OECD 2003) was commonly adopted, assuming cause-effect relationships between pressures on an environment and its response(s) to them. Indicators were enshrined in policy and are still widely used to demonstrate compliance with international, national and regional protocols and legislation. These frameworks enforced the call to 'think globally while acting locally' (United Nations Environmental Program 1992), recognising that without strategic top-down initiatives, framed at global or regional scales, competing local interests would likely compromise the capacity to achieve success, as no one would take responsibility for large-scale, gradual degradation.

Bioindicators were adopted with alacrity in the river management arena, where concerns for river health focussed on water quality problems, typically point source industrial pollutants. For example, the US introduced legislation requiring each state to biannually report the quality of its waters to the US Environmental

Protection Agency, specifying the use of bioindicators as well as other, more direct measures of contaminants (Niemi and McDonald 2004). Recognising that water quality concerns cannot be viewed independently from the physical structure of the river system itself (e.g. Graf 2001), a 'field of dreams' approach to river restoration emerged (Lake 2001). This approach is founded on the belief that biotic communities would colonise or recover if suitable physical habitat was provided. It also sought to address the habitat loss caused by interventions such as changes to channel morphology, the separation of channel and floodplain systems, and the near-complete removal of riparian vegetation and wood loadings (or their replacement by exotic species). Eventually it was realized that unless these applications built upon insights into broader ecosystem or landscape dynamics, they were doomed to fail (Postel and Richter 2003; Lepori et al. 2005; Palmer et al. 2005; Wohl et al. 2005).

Endeavours to work within an 'ecosystem' approach to environmental management recognized the need to work across disciplines. However, approaches tended to be summative, piecing together contributions from different fields of enquiry rather than developing integrated conceptual frameworks and methods (e.g. Uys 1994). Ecosystems still tended to be viewed independently from each other, with little appreciation of spatial context and connectivity (e.g. analysis of interactions among terrestrial and aquatic ecosystems remains in its infancy). The 'balance of nature' paradigm prevailed, suggesting that without human intervention, natural systems would generally evolve to a state of equilibrium. The goal of conservationists remained the restoration or preservation of this pristine state, epitomised in reference sites. The goal of environmental scientists was to understand the functional mechanics of natural systems, with much effort devoted to the development of ever more detailed process based models aiming to predict the impacts of human intervention or environmental change (e.g. eutrophication models, reviewed in Koelmans et al. 2001).

2.2.3 Chaos and Complexity: Nature as a Complex Adaptive System

Just as the ecosystem approach was being operationalized in new institutions and legislation, a dramatic shift in scientific perspective has emerged that challenges or undermines its theoretical foundation. The new paradigm in both ecology and earth sciences views natural systems as complex and adaptive, often involving non-linear relationships and stochastic events that result in effects that can be unpredictable in time and space (Phillips 1992, 2003; Perry 2002; Wallington et al. 2005; Harris 2007). It is now recognised that stability and equilibrium are merely illusions of scale (see Bracken and Wainwright 1996). Observed patterns are the cumulative result of many processes operating at many scales, in which the sequence of events and spatial configuration may be critical in determining the outcome. Furthermore, some systems may be subject to 'catastrophic shifts' in response to relatively small trigger events (Scheffer et al. 2001). A given sequence of events may potentially generate a variety of possible outcomes, such that the effects of change cannot be reliably predicted (Wallington et al. 2005).

The concept of biodiversity has widened still further. Many would now prefer the term 'biocomplexity', defined as "the multiplicity of interconnected relationships and levels" (Ascher 2001). The new term reveals its roots in systems theory, where complexity and chaos theory deal with issues such as non-linearity, self-organisation and emergence, the contingency of initial conditions and historical path dependence. Pickett et al. (2005) describe biocomplexity as having three dimensions, spatial, organizational and temporal. They argue that spatial analysis must be explicit, taking location and neighbourhood relations into account, rather than merely focusing on the number and type of entities contributing to spatial heterogeneity. Organizational complexity encompasses not only functional units, but also their connectivity, which often constrains or drives their interactions, in many cases across organizational levels. Temporal complexity reflects system evolution and legacies of past history, which may continue to impact through lagged interactions or slowly emerging indirect effects. Landscape ecology has come to the fore as two-way relationships between pattern and process are explored (Turner et al. 2001). For example, the theory of patch dynamics informs studies of the relationships between different landscape units and flows of the energy and resources vital to ecosystem function are modelled (e.g. Forman 1995; Poole 2002).

The contemporary focus is on the processes responsible for generating and maintaining observed structures and patterns. Disturbances such as fire and flood are no longer considered as aberrations, but as an integral part of system dynamics, often required for its persistence (Turner et al. 1993). The management focus has moved from conserving static ecosystems and seeking to repair damage by restoring the (definitive) 'natural' landscape, towards strategies that seek to maximise or conserve ecosystem resilience, so that systems can continue to function in the face of unexpected disturbances and human impacts (e.g. Holling 1973; Peterson et al. 1998; Gunderson 2000; Carpenter et al. 2001; Folke 2003; Walker et al. 2004). Emphasis is placed on patch boundaries and the connectivity between patches, since the maintenance of flow paths is vital to ecosystem sustainability (Forman 1995). In many systems, heterogeneity and connectivity are associated with increased system resilience, resistance to exotic invasions and the maintenance of renewal processes and ecosystem services such as nutrient recycling, pollination, detoxification and the biological control of parasites and pathogens (Hooper et al. 2005). In the river management arena, these developments are exemplified by efforts to allow river systems to self-adjust, such as the 'space to move' programmes adopted for various European rivers (Everard and Powell 2002).

Importantly, contemporary scientific thinking views humans as part of environmental systems, recognizing human needs and abandoning unrealistic assumptions that the ideal or reference state of all ecosystems is one that lacks human presence (Waltner-Toews and Kay 2005). This introduces a political dimension to determining an acceptable or desirable ecosystem state before conservation or rehabilitation goals can be set. In each landscape, contemporary local interests and cultural values need to be balanced against the sustainability of ecosystem services for use globally and by future generations (e.g. Rhoads et al. 1999). Explicit recognition of the inherent uncertainties and complexities of social-ecological systems has highlighted the need for much greater collaboration among natural and social scientists, the research community and environmental managers, management processes and the community. This is exemplified by increased emphasis on links between ecological and societal resilience, in a sense paralleling earlier relationships between ecosystem and societal health (e.g. Folke et al. 2002, 2004, 2005; Berkes et al. 2003; Dietz et al. 2003).

These considerations have prompted the emergence of a probabilistic approach to science and a precautionary, adaptive approach to decision-making and the prioritisation of issues (Ravetz 1999; Walters and Holling 1990). Realization of the non-deterministic behaviour of social-ecological systems has prompted more relaxed approaches to consideration of uncertainty, recognizing the inherent limitations of the unknown and the unknowable. For example, Palmer et al. (2005) indicate that contemporary approaches to river management should emphasize concerns for natural variability, biophysical linkages, ecosystem dynamics, resilience, uncertainty and inherent complexity, recognizing that short-term, small-scale practicalities must be framed in the context of large scale, long time scale processes. The emphasis has moved from generally applicable models designed to predict specific outcomes towards abstract simulation models and spatially explicit GIS methods that are used to suggest possible outcomes under different scenarios (e.g. Montgomery 2001; Perry 2002; Millennium Ecosystem Assessment 2005). Integrative, transdisciplinary conceptual frameworks and methods that reach across both human and physical sciences are slowly replacing the summative cross-disciplinary approach (Tress et al. 2005). These frameworks reach outside science, including both managers and stakeholders in their development and application, challenging conventional approaches to science and its uptake.

2.3 The Emergence of 'Post-Normal' Science

The 'single issue' and 'ecosystem' approaches outlined above involve the use of the traditional scientific method. This is characterised by reductionist thinking based on deductive reasoning, in which experimental methods are applied to test hypotheses that build upon established theories. Statistically rigorous procedures are used to appraise the outcomes of plot-based experiments, progressively eliminating potential explanations that are framed in terms of causal mechanisms (Popper 1959). Deterministic, quantitative relationships build on documented knowledge and are applied to develop optimum outcomes that address particular problems. The effectiveness of engineering science is testimony to our success in these endeavours. While these practices can be extremely effective in addressing

Table 2.2 Principles of normal and post-normal science

Established paradigm ('normal science')	Novel framework ('post-normal science')
Employs recognized methods:	Move beyond standard procedures:
• Deterministic	• Probabilistic—applies the precautionary approach
• Steady accumulation of relatively stable and certain facts	• Moves beyond cause and effect experimentation, recognizing multiple (non-linear) causation, using a multiple lines of evidence approach
• Experimental, deductive reasoning that focuses on elimination of potential explanations	• Recognizes the inherent variability and complexity of the natural world
Simplifies reality using a summative approach—works on parts of systems (only those deemed relevant to the target purpose), identifying core relationships which can be quantified, later piecing them together and manipulating them in management applications	Integrative, holistic approach involving systems-based conceptual models that include all parameters and interactions of importance (scientific and social)
	Merges qualitative and quantitative reasoning, placing due regard on the knowledge structures of cognate disciplines, and associated notions of precision, error, reliability and uncertainty, whilst acknowledging and incorporating local 'non-expert' knowledge
	Explicitly emphasizes broader-scale processes
Considers itself to be 'valueless' (i.e. notionally objective)	Recognizes that knowledge generation and analytical procedures are value-laden, and that interpretations vary

concerns for a particular purpose, many unwanted side effects may be experienced. For example, levee construction and dredging may create a smooth channel with sufficient flow for navigation purposes, but may compromise the integrity of aquatic ecosystems.

The emergence of 'post normal' science (Ravetz 1999) fundamentally challenges the values and methods of traditional 'normal' science (Table 2.2). By intent and design, 'post normal' enquiry addresses big-picture issues of genuine societal concern. Its agenda is negotiated through stakeholder consultation, set by pressing issues and goals, rather than by discipline-bound theories and institutions. A holistic, integrated approach to enquiry is adopted at the outset, rather than trying to piece together concepts and tools developed separately within different disciplines. Conceptual frameworks seek to organise knowledge derived from both science and experience, accepting inputs from managers and from local communities and experts alongside those provided by individual disciplines. They encompass both human and natural sciences, qualitative as well as quantitative reasoning and are validated by multiple lines of evidence, rather than by scientific proof (Downes et al. 2002).

In developing and adopting new approaches, scientists are increasingly required to go beyond their comfort-zone. Many feel uncomfortable addressing questions posed by others, particularly in situations involving high stakes and contested values, where facts and understanding are uncertain and scientists are required to make value-laden judgement calls and to incorporate lay knowledge (Funtowicz and Ravetz 1993). However, many scientists continue to consider their work within a societal vacuum. We need to acknowledge prevailing traditions, dogma and belief systems embedded in our paradigms (Kuhn 1962). The new transdisciplinary approach also challenges individuals and institutions that are accustomed to working independently, competing against each other for recognition and funding (Jakobsen et al. 2004).

2.4 Changing Approaches to Environmental Management

Without inferring a causal, linear relationships, the changing approaches to environmental science are mirrored in different styles of managerial governance and applications (Table 2.3). Under the 'command and control' approach, politically driven agendas are implemented via top-down management frameworks. Site-specific projects are framed within a decadal timeframe, using standardised solutions. They tend to have a technical focus and are typically contracted to professionally-accredited engineers with short-term commitment and accountability (e.g. Higgs 2003). Maintenance is divorced from design and construction (e.g. Williams 2001; Hillman and Brierley 2005).

The 'ecosystem' approach is also characterised by top-down management, with agendas set by professionals and little community involvement. However, the professionals are now the new class of managers established to implement the legislation and protocols designed to protect the environment (e.g. planning applications/resource consent). Indicators are widely used to set targets, often reflecting internationally or regionally set standards of acceptability. Ambitious and costly monitoring programmes are adopted, implemented by professionals. Projects still remain site-specific and tend to involve standardised solutions. Effort is devoted to the development of 'toolkits' that can be widely used (e.g. river erosion/bank stabilisation kits). Unfortunately, they are often applied with little attention to local context or expertise or the range of secondary consequences that may result. Many institutional arrangements bear the legacy of this approach, with large investment in the development of indicators and long term data sets that are difficult to abandon, even when they are no longer convinced that the measures are useful or appropriate. Even more alarming, however, was the lack of coherent management processes to clearly articulate the aims and audit the effectiveness of interventions (e.g. Bernhardt et al. 2005). Unfortunately, even when they are applied, assessment procedures that are used to appraise environmental health are often unduly prescriptive and rigid, framed largely in terms of structural measures rather than addressing broader-ranging concerns for ecosystem functionality and complexity.

Recognizing that our understanding will never be complete, adaptive and participatory management frameworks provide the most reliable basis with which

Table 2.3 Contrasting styles of environmental management

	Governance	Management applications
'Command and control' management	Top-down, politically driven Short-term commitment (and contract) with limited accountability	Applies 'average' or 'normal' solutions Site-based projects
	Monitoring is internalised or overlooked	Construction focus—typically hard engineering
	Maintenance is divorced from design	
	Extension science that 'educates' people about the environment	
'Ecosystem' management	Top-down, driven by standards enshrined in international and local legislation and protocols	Projects focus on whole ecosystems, but at reach scale only 'Toolkit' solutions applied irrespective of local context or expertise
	Indicators extensively used in ambitious monitoring programmes implemented by professionals	
	Institutions reorganised to bring experts together	
	Little community involvement or responsibility	
Adaptive management	Bottom-up, participatory—inclusive approach to visioning and prioritisation	Catchment based projects and organizations
	Long term (ongoing) commitment	Emphasize concerns for the rare, unique or culturally sensitive attributes of any given system, focussing on elements of distinctiveness
	Monitoring is externalised, placing due regard on pre/post project appraisals	Continuum of interventions, including 'do nothing', soft engineering or hard engineering options
	Maintenance is a core activity	
	Action research that promotes mutual (adaptive) learning from outcomes	

to work with, and plan for, uncertainties (Walters and Holling 1990). This approach to environmental management views people as part of nature and 'works with' the diversity and dynamics of ecosystems, aiming to restore sustainable relationships between nature and culture. Recognising a multiplicity of options, the first stage in adaptive management is to determine the desired or acceptable landscape state, balancing human and environmental needs, rather than imposing a culturally derived vision of a pristine state. This entails negotiation among conflicting interests, and associated trade-offs between sustainability and development. Participatory frameworks engage a range of stakeholders in decision-making processes. System-wide managerial actions are applied as experiments (Walters and Holling 1990), with community involvement in achievable long-term monitoring and maintenance programmes. The results feed back as learning, informing both scientific enquiry as predictions are tested and managerial practice as methods are improved. A living information-base, developed and applied within an appropriate information management system, uses best available knowledge, identifies gaps, then targets areas of deficient insight (Hillman and Brierley 2002).

2.5 Conclusion: Opportunities for Geographers in the Management of Complex Adaptive Systems

Uncertainty is a prevalent theme in the new paradigm of environmental science and management, pervading contemporary insights into ecosystem behaviour, our understanding of such phenomena, the process of knowledge transfer, and the political dimension of institutional/social applications. In this situation, scientists cannot be relied upon by managers to provide mechanistic models of nature. Outcomes are unpredictable, such that forecasting is replaced by foresighting based on probable scenarios that are constrained by the boundaries of natural variability and human actions (e.g. Millennium Ecosystem Assessment 2005). Targets are negotiable, demanding consensus between stakeholders within the limits of available tools and resources. Standard indicators cannot be used to evaluate system health and sustainability, but need to be developed in the context of local systems and human priorities. Limiting factors that constrain environmental and managerial performance must be sought out, prioritized appropriately, and addressed. Both scientists and environmental managers need to develop new ways of working, challenging established methods, norms and institutional structures (Table 2.4).

Geographers are clearly well-placed to help scientists and managers meet the challenges presented by environmental management in a changing world. Ultimately, concerns for environmental futures must build on the relationship between people and place. A healthy society underpins a healthy environment, and vice versa. People must be seen as part of nature, emphasizing concerns for the mutual interdependence of social and environmental sustainability. Opportunities abound for geographers to help meet the challenges presented in translating the new theory into sound environmental management practice. Centred on concepts of place, geographers bring skills and experience in the marriage of human and physical science that are fundamental to the successful development and implementation of adaptive management programs. Framed in a landscape context, emerging practice depends heavily on the use of modelling, GIS and remote sensing techniques frequently employed by both physical and human geographers. Human geographers have adopted leading roles in the development and implementation of the concepts of sustainable development and community participation in decision-making. Shaping ecosystem futures by working directly with managers and key decision-makers is a form of post-normal practice that lies very comfortably within Geographic discourse. Indeed, concerns for the study of

Table 2.4 Challenges to scientific and managerial practices in an uncertain and changing world (based on Rogers 2003)

Scientists must	Managers must
• Accept agendas set by real world issues, tied directly to managerial concerns and incorporating social and economic dimensions	• Negotiate objectives and measures of success agreed between stakeholders, managers and scientists, with collective ownership of outcomes
• Express opinions and 'guestimates', using arguments based on reasonable assumptions rather than scientific proof	• Accept that scientists cannot provide objective solutions, predictions or measures of success
• Communicate clearly to managers and stakeholders, sharing knowledge to promote informed choices and welcoming the contribution of local 'lay' experience	• Undertake experiments—no single course of action may emerge as 'correct'. Learning by doing turns failures into stepping stones to progress and successes can be reinforced
• Frame recommendations realistically in terms of budgets and tools available (which may be crude)	• Be flexible—situations can change as a result of natural variability, evolving theories/models, changing stakeholder people/needs/priorities
• Avoid the 'tyranny of modelling', hoping that more detailed process-based models will provide all the answers	• Develop plans and policies that can be tailored to individual locations—lessons are not easily transferred
• Develop new ways of legitimising theory in situations where the traditional scientific method is not easily applied—large scale manipulation of ecosystems is usually prohibitively costly and impractical, samples of one are not uncommon. Multiple lines of evidence can be used to support an argument.	• Reconcile concern for the long term or widespread impact of small changes with political expediency and short-termism
• Think holistically, avoiding reductionist/single purpose approaches, working across disciplines and breaking down barriers of language, theories, concepts, methods and institutions	• Cooperate between organizations, avoiding duplication of effort and recognising that the spatio-temporal domains of ecosystem processes do not necessarily match those of the authorities charged with their management

Together, they need to develop conceptual models of how ecosystems work, based on best available knowledge, using them as platforms to organise knowledge, develop treatments, prioritise management actions, develop visions and goals, predict treatment responses and review outcomes

complex socio-ecological systems present an opportunity for convergence of perspectives among physical and human geographers, building upon our traditional emphasis upon specialist-generalist syntheses and the application of divergent methodologies (both qualitative and quantitative).

For example, geomorphologists have a critical role in framing ecological considerations in terms of an appropriate understanding of the physical template upon which ecosystem processes and forms operate. Significant success has been gained in the use of conceptual frameworks with a managerial focus that apply nested-hierarchical principles to examine geomorphic relationships within catchments (e.g. Petts and Amoros 1996; Rogers and O'Keefe 2003; Brierley and Fryirs 2005). These coherent, spatially-integrated frameworks identify the distinct attributes and patterns of any given system. By extension, they can be used to characterize the key processes and biophysical relationships that maintain the integrity of the system, identifying thresholds of potential concern that may compromise the integrity of the system in the light of environmental and land use changes (Du Toit et al. 2003). When tied to appraisals of environmental history, such insights are a critical component in the assessment of the potential for ecological recovery, an essential process in maximising the efficiency and effectiveness of management programs that target environmental repair.

Adjustments to scientific practice are required as researchers are increasingly drawn into the managerial arena. Emphasis must be placed on asking the right questions, seeking integrative solutions rather than discipline-bound treatments. Ultimately, academic institutions have a significant capacity and responsibility to inform and guide long-term developments through their training of the next generation of environmental management practitioners. Curriculum reform that promotes holistic knowledge of environmental systems and management frameworks is required. Effective management of real-world problems requires a shift in focus from short-term, local-scale, discipline-bound treatments to regional-scale analyses that target particular concerns for ecosystem integrity and functionality on an ongoing basis. These must be set within a coherent framework that can unite different perspectives and methodologies. These frameworks are place-specific, recognising the unique combinations of social and biophysical features and histories that characterise each location. This repository of knowledge must be owned and shared by scientists, managers and stakeholders alike, demanding effective management and communication strategies and common, accessible language and concepts that can facilitate collaboration that moves beyond traditional silos. A proactive, future-focussed approach is needed, with reflection to identify potentially weak links that may compromise success. Weaknesses may arise from limitations in knowledge, cross-disciplinarity, transfer of knowledge (communication or misapplication), use of knowledge or institutional impediments, as well as from limitations in the community/social will, political support (vision) or available budget. However prepared we think we may be, surprising outcomes (or circumstances) are likely, so flexibility to adapt readily to new opportunities or learn from mistakes must be retained. Ultimately, prospects for successful environmental repair are contingent upon effective societal engagement in the setting of environmental goals and the application of measures to address them—regardless of the quality of scientific understanding with which we may aspire to address these issues.

Acknowledgments We thank Susan Owen, Brad Coombes and Mick Hillman for helpful comments in the development of this manuscript, and David Higgitt for his co-ordination of support to present an earlier version of this work at the National University of Singapore Centenary Symposium.

References

Ascher W (2001) Coping with complexity and organizational interests in natural resource management. Ecosystems 4:742–757

Berkes F, Colding J, Folke C (eds) (2003) Navigating social—ecological systems: building resilience for complexity and change. Cambridge University Press, Cambridge

Bernhardt ES, Palmer MA, Allan JD, Alexander G, Barnas K, Brooks S, Carr J, Clayton S, Dahm C, Follstad-Shah J, Galat D, Gloss S, Goodwin P, Hart D, Hassett B, Jenkinson R, Katz S, Kondolf GM, Lake PS, Lave R, Meyer JL, O'Donnell TK, Pagano L, Powell B, Sudduth E (2005) Synthesizing US river restoration efforts. Science 308:636–637

Block WM, Brennan LA, Gutierrez RJ (1987) Evaluation of guild-indicator species for use in resource management. Environ Manag 11:265–269

Botkin DB (1989) Discordant harmonies: a new ecology for the twenty-first century. Oxford University Press, Oxford

Bracken LJ, Wainwright J (2006) Geomorphological equilibrium: myth and metaphor? Trans Inst Br Geogr NS31:167–178

Brierley GJ, Fryirs KA (2005) Geomorphology and river management: applications of the river styles framework. Blackwell, Malden

Carpenter SR, Walker B, Anderies JM, Abel N (2001) From metaphor to measurement: resilience of what to what? Ecosystems 4:765–781

Claridge MF, Dawah HA, Wilson MRE (1997) Species: the units of biodiversity. Chapman & Hall, London

Cosgrove D, Petts GE (eds) (1990) Water, engineering and landscape. Belhaven Press, London

Dietz T, Ostrom E, Stern PC (2003) The struggle to govern the commons. Science 302:1907–1912

Downes BJ, Barmuta LA, Fairweather PG, Faith DP, Keough MJ, Lake PS, Mapstone BD, Quinn GP (2002) Monitoring ecological impacts: concepts and practice in flowing waters. Cambridge University Press, Cambridge

Du Toit JT, Rogers KH, Biggs HC (eds) (2003) The Kruger experience: ecology and management of savanna heterogeneity. Island Press, Washington DC

Everard M, Powell A (2002) Rivers as living systems. Aquat Conserv-Mar Freshw Ecosyst 12:329–337

Folke C (2003) Freshwater and resilience: a shift in perspective. Philos Trans Royal Soc London Ser B 358:2027–2036

Folke C, Carpenter SR, Elmqvist T, Gunderson L, Holling CS, Walker B (2002) Resilience and sustainable development: building adaptive capacity in a world of transformations. Ambio 31:437–440

Folke C, Carpenter S, Walker B, Scheffer M, Elmqvist T, Gunderson L, Holling CS (2004) Regime shifts, resilience, and biodiversity in ecosystem management. Annu Rev Ecol Evol Syst 35:557–581

Folke C, Hahn T, Olsson P, Norberg J (2005) Adaptive governance of social-ecological systems. Ann Rev Environ Resour 30:441–473

Forman RT (1995) Land mosaics. The ecology of landscapes and regions. Cambridge University Press, Cambridge

Franklin JF (1993) Preserving biodiversity: species, ecosystems, or landscapes. Ecol Appl 3:202–205

Funtowicz SO, Ravetz JR (1993) Science for the post-normal age. Futures 25:739–755

Gaston KJ (1996) What is biodiversity? In: Gaston KJ (ed) Biodiversity: a biology of numbers and difference. Blackwell, Oxford, pp 1–9

Graf WL (2001) *Damage* control: restoring the physical integrity of America's rivers. Ann Assoc Am Geogr 91(1):1–27

Gunderson LH (2000) Ecological resilience: in theory and application. Annu Rev Ecol Syst 31:425–439

Harris G (2007) Seeking sustainability in an age of complexity. Cambridge University Press, New York

Heino J, Paavola R, Virtanen R, Muotka T (2005) Searching for biodiversity indicators in running waters: do bryophytes, macroinvertebrates, and fish show congruent diversity patterns? Biodivers Conserv 14:415–428

Higgs E (2003) Nature by design: people natural process and ecological restoration. MIT Press, Cambridge

Hilderbrand RH, Watts AC, Randle AM (2005) The myths of restoration ecology. Ecol Soc 10(1):19 [online]

Hillman M, Brierley GJ (2002) Information needs for environmental flow allocation: a case study from the Lachlan River, New South Wales, Australia. Ann Assoc Am Geogr 92:617–630

Hillman M, Brierley GJ (2005) A critical review of catchment-scale stream rehabilitation programmes. Prog Phys Geogr 29:50–70

Holling CS (1973) Resilience and stability of ecological systems. Annu Rev Ecol Syst 4:1–23

Holling CS, Meffe GK (1996) Command and control and the pathology of natural resource management. Conserv Biol 10:328–337

Hooper DU, Chapin FS, Ewel JJ, Hector A, Inchausti P, Lavorel S, Lawton JH, Lodge DM, Loreau M, Naeem S, Schmid B, Setala H, Symstad AJ, Vandermeer J, Wardle DA (2005) Effects of biodiversity on ecosystem functioning: a consensus of current knowledge. Ecol Monogr 75:3–35

Jakobsen CH, Hels T, McLaughlin WJ (2004) Barriers and facilitators to integration among scientists in transdisciplinary landscape analyses: a cross-country comparison. For Policy Econ 6:15–31

Jones CG, Lawton JH, Shachak M (1994) Organisms as ecosystem engineers. Oikos 69:373–386

Kay JJ, Regier HA, Boyle M, Francis G (1999) An ecosystem approach for sustainability: addressing the challenge of complexity. Futures 31:721–742

Koelmans AA, van der Heijde A, Knijff LM, Aalderink RH (2001) Integrated modelling of eutrophication and organic contaminant fate and effects in aquatic ecosystems. A review. Water Res 35:3517–3536

Kuhn TS (1962) The structure of scientific revolutions. University of Chicago Press, Chicago

Lake PS (2001) On the maturing of restoration: linking ecological research and restoration. Ecol Manag Restor 2:110–115

Lambeck RJ (1997) Focal species: a multi-species umbrella for nature conservation. Conserv Biol 11:849–856

Launer AE, Murphy DD (1994) Umbrella species and the conservation of habitat fragments: a case study of a threatened butterfly and a vanishing grassland ecosystem. Biol Conserv 69:145–153

Lepori F, Palm D, Brannas E, Malmqvist B (2005) Does restoration of structural heterogeneity in streams enhance fish and macroinvertebrate diversity? Ecol Appl 15:2060–2071

Losos E (1993) The future of the US endangered species act. Trends Ecol Evol 8:332–336

Millennium Ecosystem Assessment (2005) Ecosystems and human well-being: synthesis. Island Press, Washington DC

Montgomery D (2001) Geomorphology, river ecology, and ecosystem management. In: Dorava JM, Montgomery DR, Palcsak BB, Fitzpatrick FA (eds) Geomorphic processes and riverine habitat. American Geophysical Union, Washington DC, pp 247–253

Niemi GJ, McDonald ME (2004) Application of ecological indicators. Annu Rev Ecol Evol Syst 35:89–111

Noss RF (1990) Indicators for monitoring biodiversity: a hierarchical approach. Conserv Biol 4:355–364

OECD (2003) Environmental indicators: development, measurement and use. www.oecd.org/dataoecd/20/40/37551205.pdf. Accessed Jan 2011

O'Neill RV (2001) Is it time to bury the ecosystem concept? (With full military honors of course!). Ecology 82:3275–3284

Paine RT (1995) A conversation on refining the concept of keystone species. Conserv Biol 9:962–964

Palmer MA, Bernhardt ES, Allan JD, Lake PS, Alexander G, Brooks S, Carr J, Clayton S, Dahm CN, Shah JF, Galat DL, Loss SG, Goodwin P, Hart DD, Hassett B, Jenkinson R, Kondolf GM,

Lave R, Meyer JL, O'Donnell TK, Pagano L, Sudduth E (2005) Standards for ecologically successful river restoration. J Appl Ecol 42:208–217

Pearson DL (1994) Selecting indicator taxa for the quantitative assessment of biodiversity. Philos Trans Royal Soc London Ser B-Biolog Sci 345:75–79

Perry GLW (2002) Landscapes, space and equilibrium: shifting viewpoints. Prog Phys Geogr 26:339–359

Peterson GD, Allen CR, Holling CS (1998) Ecological resilience, biodiversity, and scale. Ecosystems 1:6–18

Petts GE, Amoros C (eds) (1996) Fluvial hydrosystems. Chapman & Hall, London

Phillips JD (1992) Nonlinear dynamical systems in geomorphology: revolution or evolution? Geomorphology 5:219–229

Phillips JD (2003) Sources of nonlinearity and complexity in geomorphic systems. Prog Phys Geogr 27:1–23

Pickett STA, Cadenasso ML, Grove JM (2005) Biocomplexity in coupled natural-human systems: a multidimensional framework. Ecosystems 8:225–232

Poole GC (2002) Fluvial landscape ecology: addressing uniqueness within the river discontinuum. Freshw Biol 47:641–660

Popper KR (1959) The logic of scientific discovery. Hutchinson, London

Postel S, Richter B (2003) Rivers for life Managing water for people and nature. Island Press, Washington DC

Prendergast JR (1997) Species richness covariance in higher taxa: Empirical tests of the biodiversity indicator concept. Ecography 20:210–216

Ravetz JR (1999) What is post-normal science? Futures 31:647–653

Rhoads BL, Wilson D, Urban M, Herricks EE (1999) Interaction between scientists and nonscientists in community-based watershed management: Emergence of the concept of stream naturalization. Environ Manag 24:297–308

Richter B, Postel S (2004) Saving earth's rivers. Issues Sci Technol 20:31–36

Ricotta C (2005) Through the jungle of biological diversity. Acta Biotheor 53:29–38

Rogers KH (2003) Adopting a heterogeneity paradigm: Implications for management of protected savannas. In: Du Toit JT, Rogers K, Biggs HC (eds) The Kruger experience Ecology and management of savanna heterogeneity. Island Press, Washington DC, pp 41–58

Rogers KH, O'Keefe J (2003) River heterogeneity: ecosystem structure, function and management. In: Du Toit JT, Rogers K, Biggs HC (eds) The Kruger experience Ecology and management of savanna heterogeneity. Island Press, Washington DC, pp 189–218

Scheffer M, Carpenter S, Foley JA, Folke C, Walker B (2001) Catastrophic shifts in ecosystems. Nature 413:591–596

Severinghaus WD (1981) Guild theory development as a mechanism for assessing environmental-impact. Environ Manage 5:187–190

Simberloff D (1998) Flagships, umbrellas, and keystones: Is single-species management passe in the landscape era? Biol Conserv 83:247–257

Sprugel DG (1991) Disturbance, equilibrium, and environmental variability: what is natural vegetation in a changing environment? Biol Conserv 58:1–18

Tress G, Tress B, Fry G (2005) Clarifying integrative research concepts in landscape ecology. Landsc Ecol 20:479–493

Turner MG, Romme WH, Gardner RH, O'Neill RV, Kratz TK (1993) A revised concept of landscape equilibrium: disturbance and stability on scaled landscapes. Landsc Ecol 8:213–227

Turner M, Gardner RH, O'Neill RV (2001) Landscape ecology in theory and practice: pattern and process. Springer, New York

United Nations Environmental Program (1992) Convention on Biological Diversity. http://www.cbd.int/. Accessed Jan 2011

Uys M (ed) (1994) Classification of rivers and environmental health indicators. In: Proceedings of a Joint South African/Australian workshop, February 7–14 1994, Cape Town, South Africa. Water Research Commission Report No. TT 63/94

Walker BH, Holling CS, Carpenter SR, Kinzig AS (2004) Resilience, adaptability and transformability in social-ecological systems. Ecol Soc 9(2):5 [online]

Wallington TJ, Hobbs RJ, Moore SA (2005) Implications of current ecological thinking for biodiversity conservation: a review of the salient issues. Ecol Soc 10(1):15 [online]

Walters CJ, Holling CS (1990) Large-scale management experiments and learning by doing. Ecology 71:2060–2068

Waltner-Toews D, Kay J (2005) The evolution of an ecosystem approach: the diamond schematic and an adaptive methodology for ecosystem sustainability and health. Ecol Soc 10(1):38 [online]

Wegner A, Moore SA, Bailey J (2005) Consideration of biodiversity in environmental impact assessment in Western Australia: practitioner perceptions. Environ Impact Assess Rev 25:143–162

Williams PB (2001) River engineering versus river restoration, ASCE Wetlands Engineering and River Restoration Conference, Reno, NV

Wohl E, Angermeier PL, Bledsoe B, Kondolf GM, Macdonnell L, Merritt DM, Palmer MA, Poff NL, Tarboton D (2005) River restoration. Water Resour Res 41(10):W10301

Wu JG, Loucks OL (1995) From balance of nature to hierarchical patch dynamics: a paradigm shift in ecology. Q Rev Biol 70:439–466

Chapter 3
River Hardware and Software: Perspectives on National Interest and Water Governance in the Mekong River Basin

Philip Hirsch

Abstract At a global level, river basin development and management has shifted from a 'hardware'-driven approach based around engineering river systems in the form of dams, diversions and other large structures, toward a 'software'-driven approach under the broad rubrics of governance and integrated water resource management. Nevertheless, large-scale water resource development is still being pushed ahead. There is clearly not an 'either/or' scenario in terms of hardware and software approaches to river management. This chapter examines the implications of new approaches to river basin governance for the planning and implementation of river engineering structures in a transboundary river setting. The context for the study is the Mekong river basin. The Mekong has achieved prominence among the world's more than 260 river basins that cross national boundaries, as a river and a basin that is actively managed across borders. One of the reasons for such prominence is the established institutional basis for cooperation among the four lower countries of the basin and the international support for this governance framework. Another is the longstanding and continuing plans for significant impoundment and diversion of the river and its tributaries. At present, the Mekong is moving toward something of a crisis of transboundary water governance. The Mekong River Commission (MRC) is at the heart of this crisis. At one level, the conundrum is the tension between management of the river for ecological sustainability and social justice, on the one hand, and the drive for development of a relatively under-exploited set of water resources on the other. This tension is exaggerated in a river basin whose population remains economically poor and heavily dependent on the natural resource base for livelihood. At another level, the conundrum is one of scale of governance, and this poses both challenges and opportunities for the MRC as an integrated water resource management agency.

P. Hirsch (✉)
Australian Mekong Resource Centre, School of Geosciences (F09),
University of Sydney, Sydney 2006, Australia
e-mail: philip.hirsch@sydney.edu.au

D. Higgitt (ed.), *Perspectives on Environmental Management and Technology in Asian River Basins*, SpringerBriefs in Geography, DOI: 10.1007/978-94-007-2330-6_3, © The Author(s) 2012

Keywords Water governance · Transboundary issues · Integrated water resources management · Mekong river commission · Scaling issues

3.1 Introduction: River Basin Hardware and Software

River basin development has long been associated with large scale infrastructure, notably hydropower facilities and irrigation systems. River basin authorities have been established to develop river basin systems in a coordinated way through the planning of interconnected complexes of storages, diversions and delivery systems, initially inspired by the Tennessee Valley Authority (Ekbladh 2002; Svendsen et al. 2005). Development and subsequent management of water resources in their river basin context has long been treated as a modernist project of 'hardware' development in the form of concrete structures that impound, divert and store water for energy, irrigation and flood control purposes.

Problems associated with overdevelopment of river basins have become increasingly manifest at both project and river basin levels. Large dams have well-documented environmental and social impacts, and the unequal geographical distribution of costs, benefits and risks that these impacts impose is recognized at scientific and societal levels (McCully 1996; World Commission on Dams 2000). Over-allocation of water in river basins such as the Murray-Darling, and declining water quality in basins such as the Rhine, have forced a recognition that water needs to be managed socially and environmentally as well as technologically (Shah et al. 2003).

Increasingly, the response has been to see river basin management as a regulatory function based around allocation of water as a limited, finite good within interconnected systems. Interconnectivity refers to the physical, ecological and societal connections that mean that development or abstraction of water resources in one part of a river system leads to impacts and flow-on effects in other parts. The significance of interconnectivity increases as basins progressively 'close' with commitment of ever greater proportions of available water to human or recognized environmental needs and functions. That is, in 'open' basins where large amounts of water remain uncommitted for maintaining key human and ecological functions, the connections between actions in one part of a basin and impacts in other parts are not felt so keenly, whereas more complete commitment of water to maintaining key functions in 'closed' basins means that any further subtraction or human-induced hydrological change has more keenly felt third-party effects elsewhere within the basin (Molle et al. 2006).

Understanding river basins as integrated natural systems is matched by the growing role of water in shaping social, political and economic relations, and this brings questions of rights, values and determining stakeholder preferences around water under the spotlight. In other words, attention has turned to the governance of water in river basins, and the technology of management has shifted to the 'software' of social, economic and political means to allocate and use resources equitably, efficiently and sustainably. By 'software' I refer to the range of regulatory practices, institutional forms, norms and other aspects of what increasingly comes under the rubric of 'water governance'.

Large river basins pose particular governance challenges. The mismatch between river basin and administrative boundaries immediately poses a transboundary problem. One response has been to establish river basin organizations with jurisdictions that transcend administrative-political boundaries. Historically, it has been the exception rather than the rule that international river basins have been governed by a transnational authority.

Institutional aspects of governance have been accompanied by development of 'soft' technologies. Integrated water resources management (IWRM), and its subset of integrated river basin management (IRBM) has emerged as a holistic governance response that has become increasingly mainstream and institutionalized, yet largely unproven (Biswas et al. 2005). With the adoption of IWRM and the emergence of a new orthodoxy, or 'water consensus', new questions are being asked and challenges posed to the notion of an all-encompassing framework for water management (Franks 2004).

In this chapter, I explore the implications of river basin governance in the Mekong, a transboundary river basin that has seen a shift from hardware to software approaches as formulated above, and one that has also seen both hardware and software-focused technology transfer from the US and Australia. The chapter shows that managing transboundary interests is more complex than negotiating one riparian nation's allocation against another through a single river basin authority. Scale issues are fundamental. Just as there are qualitative as well as quantitative differences between the large scale hardware of mega-dams and smaller structures, so governance is scaled in a way that has a significant bearing on outcomes and which poses significant challenges in a transboundary river basin such as the Mekong.

3.2 Scales of Governance in the Mekong

"If the wars of this century were fought over oil, the wars of the next century will be fought over water" (World Bank Vice President: Ismail Serageldin, 1995).

The prospect that nations might increasingly resort to war in order to secure access to fresh water has given immediacy to the need for effective mechanisms to manage rivers that flow across national boundaries. More than 260 of the world's rivers drain from territory in more than one country, and about 45% of the world's land surface is located in transboundary river basins. Only a small number of transboundary river basins have institutionalized governance frameworks, and most of these have been established quite recently.

Not surprisingly, the Mekong attracts interest and international funding support as a transboundary river that has been 'governed' institutionally for the more than half a century. Since the establishment of the Mekong Committee in 1957, and its reincarnation as the Mekong River Commission (MRC) in 1995, the four lower countries have had a basis for cooperation in the use and management of the Mekong. Furthermore, this period has seen both geopolitical and ecopolitical shifts that have effected fundamental changes in the primary concerns and modus operandi if the Committee/Commission. In particular, the influence of the US Army Corps of Engineers and Bureau of Reclamation, which set a leading agenda of hardware in the form of large dams from the 1950s to 1980s, has given way to a more holistic and process-oriented river basin management approach under the rubric of integrated river basin management and integrated water resources management.

While much has changed with the times in the Mekong transboundary governance framework, there is continuity in the basic conceptualization of the MRC as an institution effecting cooperation and compromises between its sovereign member states. The water wars discourse is never very far away, particularly in a region that has quite recently emerged from geopolitical conflict embroiling the four MRC member states of Thailand, Laos, Cambodia and Vietnam, and in a set of regional institutions such as MRC or the Asian Development Bank's Greater Mekong Subregion program whose broader development agenda is sometimes framed in terms of "reaping the peace dividend" (Pante 1996). In other words, the resolution of potentially conflicting 'national interests' of member and non-member riparian states remains the concern for governing the river basin. The need for MRC to work within, and subservient to, the national interest dictum and the national sovereignty paradigm remains under-explored, in part because the notion of a transboundary framework is sometimes assumed to transcend the parochial interests of nation states and to foster peace through development.

Critique of the water wars discourse has taken a number of forms. One is to show that cooperation between countries that share international rivers is much more common than conflict (Wolf 1999). However, this level of critique retains the notion of sovereign nation states as the principal level of analysis and of commonality of interest. At another level, the critique is a Mekong-specific application of calls to recognize the limitations of state-centric approaches and monolithic constructions of interests at country level. The critique is informed by North American and European experience of increasingly complex arrays of interests that are less than ever defined and self-identified in national terms (Blatter and Ingram 1998).

To pare apart the national interest question in transboundary river basin management, it is worth returning briefly to the 'water wars' metaphor. To date, while there have certainly been instances of international tension and conflict over water in the Middle East and elsewhere, the dominant axes and scales of tension over development of water resources have been socially constructed within countries. To the extent that they have taken a transboundary dimension, water conflicts are nevertheless often between different sorts of interests in the countries concerned rather than between sovereign governments. There is a need to rescale our concern over water conflict toward the more nuanced and socially framed tensions that shape the ways in which we treat our rivers and compete for their resources—for example by considering whether the appropriate metaphor is 'water wars' or 'water riots' (Boesen and Ravnborg 2003). In the Mekong region, the most significant conflicts over dams, for example, have been in protests since the 1980s in Thailand that are largely internal to that country, rather than in open dispute between riparian governments over claims to the Mekong's waters. Shiva (2002) expresses this ideologically as a question of 'paradigm wars' rather than water wars in the conventional security sense.

In the next section, I raise the question of whether the Mekong currently faces something of a crisis of governance—not because the countries that share the river are about to start shooting at one another, but rather because the different visions for the Mekong are less than ever predicated on differences between one national government and another. Yet the MRC as the main governance body, a decade since its establishment, still works largely within governance arrangements in which riparian representation is at a national level and in which development of the Basin is the shared point of interest. MRC consequently faces some fundamental choices in terms of direction and in terms of what sort of river basin governance arrangements it works within.

3.3 A Crisis of Governance?

The Mekong faces an unprecedented set of development pressures. Hydropower development on the upper mainstream and tributary rivers has received a new lease of life. The 2004 go ahead for Nam Theun two dam by the World Bank board opened the way for many more such projects to be funded after a lull of more than a decade in Bank funding for large dams. Through its Mekong Water Resources Assistance Program, the World Bank sought to facilitate a new wave of investment in water resource infrastructure in the Basin, but packaged within the framework of IWRM. Increasing scarcity of water in Thailand, caused by increase in demand as much as by natural conditions, has created calls for water transfers and irrigation development, especially in the country's dry northeastern region.

Meanwhile, non-governmental organizations that have opened spaces for civil society influence over decisions on how to manage water find themselves in a relatively weak position. Within the Mekong, Thai NGOs have carved themselves more space than those in neighbouring countries, but their role has been largely reactive against large dams. Their influence diminished somewhat under the Thaksin government. Elsewhere in the region, non-governmental voices have shown increasing confidence, so that in Cambodia and in China, for example, we see increased questioning of dams and other river basin development projects. In Laos and Vietnam, NGOs remain weak, although some alternative civil society voices have started to make themselves heard in Vietnam. This differential civil society space results in an unequal expression of concerns and interests within the basin (Hirsch 2001).

Bureaucratic governance of water varies from one national context to another. In Thailand, the establishment of 29 river basin committees in the country's 25 main river basins can be interpreted as a decentralization of authority, a shift from administratively to bio-regionally defined boundaries in water management, and an opportunity for non-governmental representation on the Committees with authority to secure resources and make decisions. Alternatively, it can be seen as an extension downward of bureaucratic authority via the recently established Department of Water Resources in the Ministry of Natural Resources and Environment. More realistically, the RBCs contain elements of both centralization and decentralization, participation and bureaucratization, and are better seen as one of a number of arenas for negotiation and evolution of governance frameworks (Thomas 2005). They are another level of river basin 'software'. RBCs are also being established in other Mekong countries, partly with the support of the Asian Development Bank and linked through a Network of Asian River Basin Organizations.

At the transboundary level, however, the MRC is the primary framework for river basin governance and it is here that the dilemmas of governance are most pronounced. The dilemmas are posed in a number of ways. The first is that of how a transboundary river basin organization operates when it does not include the upper two riparian states, including the most powerful and the most active in terms of water resource development (Kliot et al. 2001). China's ability to act unilaterally is based on its upstream location, its power vis-à-vis smaller downstream countries and its non-membership of MRC, hence non-subscription to the existing procedures or any rules that may eventuate. On the other hand, brakes are put on China's river basin development, albeit applied somewhat lightly, by increasing internal debate and concern over large scale hydropower projects and a growing environmental movement, by China's concerns to maintain good relations with ASEAN and individual Mekong countries, and by its observer status at MRC.

The second dilemma relates to the governance structure and ownership of the Commission itself. The Mekong Council as the political arm and the Joint Committee as the bureaucratic governing arm of MRC represent riparian governments as the MRC's decision-making bodies. As such, they articulate each country's position via key ministries, most of which have a strong bias toward water resource development agendas in the respective countries. MRC thus allows an undifferentiated 'national interest' of each riparian member to obfuscate the complex and contested interests in river basin management and water resource development that exist within each country and sometimes transcend national borders. The third arm of MRC, the Secretariat, is a heavily donor-dependant entity that has made significant progress toward open-ness, understanding the river ecology through river science, incorporation of local knowledge in fisheries and other basin-wide programs, the bringing of international experience in holistic approaches to river basin management and an emphasis on process in development planning and the setting of procedures based on good river science. Finally, the National Mekong Committees serve an important governance role as intermediaries between MRC and interests internal to each riparian member state. To date, however, NMCs remain poorly articulated with civil society or even with ministries and departments beyond those most immediately involved in water resource development. There is a dislocation between these levels of governance, and a potential dysfunctionality as a result.

A third dilemma concerns the nature of the river basin institution. What sort of river basin institution does MRC aspire to be? At least five potential roles can be identified or postulated (Hirsch et al. 2006). These are:

- MRC as a regulatory agency, responsible for setting rules by which riparian member states can be held to the spirit and the letter of the 1995 Agreement
- MRC as a multi-stakeholder forum that gives a platform diverse river basin users and managers to assert and negotiate their interests
- MRC as a scientific agency for decision support, so that decisions to develop water or other resources in the basin can be taken in cognizance of ecologically and socially complex transboundary impacts
- MRC as an investment broker, based on the relative underdevelopment of the Basin's water resources and the perception that attracting investment for infrastructure to increase energy, irrigation and navigation potential can improve the lives of those who live in the Basin
- MRC as a planning body, which takes on all the above roles to provide a sustainable planning framework based on a model of rational decision-making and an assumption that MRC is a governing rather than a governed body, able to implement and enforce plans form Basin to national to local levels

Each of these roles is somewhat different. They are not necessarily mutually incompatible, but there are clear tensions between them. The leadership of the Mekong Secretariat until early 2008, for example, placed heavy emphasis on MRC as an investment broker, potentially short-circuiting the more patient process-oriented approach to planning, participation and application of good science to development decisions that has been built up in recent years.

All these governance dilemmas hinge in part on the fact that MRC is predicated on assimilating the national interests of riparian nation states into basin-wide developmental interests. Such an approach to transboundary governance misses key dimensions of commonality and divergence of interest at various scales.

3.4 Three Cases

The challenges for MRC in taking on a governance role are indicated through most of the principal instances of current and recent water resource development in the Basin. Each case presented below represents countries pushing ahead with river hardware for benefits that are largely captured within their own territories, revealing the limitations of governance as it is conceived in terms of managing national interests toward a basin-wide common good. The first case shows how MRC's requirement to operate through formal national channels blocks the access of downstream communities (in Cambodia) affected by upstream hydropower development (in Vietnam) on a major Mekong tributary. The second case shows how specific interests of a non-member state (China) are legitimized discursively by reference to the benefits of downstream countries. The third case shows how multilateral MRC rules are potentially side-stepped by bilateralism to allow a major water infrastructure project to go ahead in Thailand that has significant transboundary components and potential impacts.

3.4.1 Bilateral Relations Along a Tributary: The Sesan in Vietnam and Cambodia

Hydropower development on the upper Sesan and Srepok Rivers in Vietnam's Central Highlands have potentially major impacts on downstream river ecology and indigenous minority riparian communities in Cambodia. This potential has been realized in the case of Yali Falls Dam, to date the second-largest hydropower project in the lower Mekong Basin (i.e. the area within the jurisdiction of MRC). Flooding and fluctuation in river levels along the Sesan River in Cambodia has affected up to 50,000 people in 90 communities in Ratanakiri and Stung Treng provinces and has directly led to drownings, flooding of agricultural land, loss of boats, nets and river bank gardens among other impacts. Loss of life, livelihoods and property has not been compensated or even properly enumerated by those responsible for the dam, yet further dams have been built on the Sesan and Srepok tributaries in the Central Highland with continuing downstream impacts.

Elsewhere, Hirsch and Wyatt (2004) have documented in more detail and synthesized more systematically the downstream impacts of Yali Falls dam and associated hydrological changes, the response of affected communities in trying to scale up their concerns through MRC as a transboundary institutional channel, and the ways in which such efforts have been obstructed by the government-to-government framework invoked by MRC that in effect marginalizes indigenous community voices. MRC is precluded from responding directly to community or civil society concerns and can only act on the request of national governments. Those governments are limited in terms of what the relevant ministries do and do not want to bring to the table, and in their preparedness to represent the interests of marginalized voices and ecological impacts within their own territories. Riparian governmental interaction is also embedded within the realpolitik of wider government-to-government relations.

3.4.2 Bearing the Burden: Chinese Dams and Basin Interests

The greatest alteration to Mekong hydrology currently underway is the damming of the Lancang Jiang (upper Mekong) river in Yunnan Province of southwestern China (McCormack 2001). Eight dams are being built on the mainstream, of which four are complete and two under construction, including one that is the equal tallest dam in the world at nearly 300 m (Xiaowan Dam). Dam construction impacts are exacerbated downstream by China's blasting of rapids to provide for navigation by ever larger boats to enhance trade with Thailand.

He Daming suggests that the Chinese dams on the Lancang Jiang will help downstream countries in many ways. The hydrological argument is quite a simple one, and has been articulated clearly in Plinston and He (1999). The Mekong is a monsoonal river, with pronounced flood peaks and periods of low flow. Storage dams in Yunnan can reduce the flood peak and enhance dry season flow, assisting dry season irrigation development downstream in the Delta—and perhaps in northeastern Thailand if the Khong-Chi-Mun diversion is realized in its entirety— and mitigating the devastating floods such as those which annually take the lives of several hundred people in Vietnam.

This argument assumes that seasonal fluctuation in river flows is a 'problem'. It fails to take into account the role of floods in supporting the fishery that provides two to three million tons of animal protein per annum to help feed the Basin's poor, or the role of the flood in maintaining the ecology and vital ecosystem services of the Tonle Sap in Cambodia. However, it is brought to bear to attenuate the sense that China is acting irresponsibly or selfishly as an upstream country.

More recently, He and Chen (2002) have brought a more sophisticated argument to bear. The Mekong regional economies are growing. Economic growth means growth in energy demand. Dams are inevitable. The number of people displaced from their homes, and the area of land flooded per megawatt of electricity generated is much greater in flatter areas of the lower basin than in the steep gorges of Yunnan. Rational planning suggests that it makes more sense to build dams in China rather than in the lower countries. China is willing to bear that burden and will resettle displaced people in urban areas to avoid the environmental impacts and social hardship that result from poorly planned agricultural resettlement schemes.

Framing the argument for the Lancang Jiang dams in such a way employs a rationality which suggests that there is a basin-wide interest to be served by concentrating the response to regional energy demand in China, optimizing socially, financially and ecologically by invoking a minimal cost per unit of energy. The question of who bears the burden within China remains a moot point. Increasingly, however, alternative voices have emerged (Yang 2004), although periodic difficulties faced by NGOs such as Green Watershed remind us of the limits to official toleration of the airing of such internal differences beyond the country's borders (Hirsch 2001).

3.4.3 Sustainably Integrated Water Resources Management: The Thai Water Grid

In Thailand, recent droughts have reinforced the push for a Water Grid that links 'water surplus' river basins with 'water deficit' basins. The scheme is predicated on a series of inter- and intra-basin diversion schemes, framed within the governance structures of the Department of Water Resources, most notably the River Basin Committees that employ basin boundaries as the unit for management. However, rather than using such basin organizations to manage water as a limited resource within each of these bioregions, this 200 billion baht (USD$5 billion) project is predicated on making up shortages by constructing infrastructure that will allow re-balancing natural water surpluses and deficits through a system of pumps, pipes, tunnels and canals. In this case, the software-derived discourse of IWRM is being employed to make the case for hardware development in the form of a physically-integrated water grid.

The official name of the Water Grid project is the "Sustainably integrated water management project (*Khroongkaan jadkaan nam baeb buranakaan yaang yangyeun*)". Integrated management in this case is interpreted as a means to integrate water movements much as a national or international energy grid or telephone system would channel network flows. Of course, such a scheme is not unique to Thailand. Historically, the Snowy Mountains scheme in Australia, California's north–south water carrier system, and more recently proposals for river linkage in India and China's South–North diversion from the Yangtze to Yellow River systems all follow a similar supply-based logic to dealing with water shortage.

While the Water Grid is a Thai-based scheme, it has an important international component. The project entails importing water from Cambodia (Stung Nam), Myanmar (Salween tributaries), Thai tributaries that flow into the international Mekong River (Kok and Ing), and Laos. Importation of water from Laos into Isan is internal to the Mekong River Basin. However, rather than extracting water from the Mekong River itself, the water is to be siphoned from the Nam Ngum and Xe Banghieng tributaries under the Mekong and transported to reservoirs on the Isan side. The logic of this is that extraction from the Mekong would require multi-lateral agreement through MRC. Buying water from tributaries located entirely within Laos makes the scheme a bilateral issue over which the MRC framework seems to have no jurisdiction. Whether this is indeed the case depends on interpretation of the 1995 *Agreement on the Cooperation for the Sustainable Development of the Mekong River Basin*, but if and when the necessary bilateral water trading agreement is put in place, it would require one of the other MRC riparian members to issue a challenge (Hirsch et al. 2006).

3.5 Beyond National Interest in Transboundary River Basin Management

All three cases described above demonstrate the limits to national interests as a basis for negotiating and managing water across boundaries in the Mekong and other transboundary river basins. Most tension and conflict over water is felt and enacted at intra-societal levels, involving complex interactions between state and civil society actors, between infrastructure developers and affected peoples, between public and private interests, between different sectors, and sometimes between neighbouring households and communities.

Where does this place transboundary governance, the 'software' of river management? Through MRC, it would appear that there are many projects on which the Mekong Council and Joint Committee can cooperate. Yet representatives of different countries' water ministries, for example, may have more in common than those same representatives versus fishers affected by a dam. With the legacy of MRC in the Mekong Committee as an agency whose agenda was set in developmentalist project formulation and whose mediation was between states rather than societies, 'hardware' is still high on the political agenda and in the understanding of some riparian government actors' priorities for the Commission. This should not blind us to the very significant challenges raised by the transboundary reality of the Mekong. However, it may be more useful to see this as a reinforcing challenge rather than a defining one.

The case studies above reveal the limits to transboundary governance when it occurs through a not very transparent or representative governance process based around assumed national interests. The difficulty faced by indigenous Sesan communities in trying to access dam developers in another national space, for example, creates an added complexity in managing stakeholder interests within this tributary basin.

On the other hand, and perhaps more optimistically in looking beyond MRC at the bigger picture of transboundary governance, such a governance framework risks missing an increasingly variegated set of relations and commonalities of interests across borders at other levels. For example, increasingly critical voices within China are running arguments parallel to civil society groups in Thailand, suggesting that the main arrays of interests transcend national borders. Numerous non-governmental, academic and other networks and interactions also provide at least embryonic transboundary 'governance from below'. The sidestepping of multilateral rules over transboundary water sharing so that Thailand and Laos can make bilateral arrangements, with the same hydrological effect as extracting water from the Mekong, suggests that the MRC is simply not in a position to achieve its vision to use its Basin Development Plan as an encompassing regulatory investment framework. Rather, finer scales of resolution are needed where water is not simply managed as belonging to one country or another, but is rather seen as negotiable among different configurations of stakeholders. Only by engaging at such levels would MRC have a more holistic role in governance terms.

The governance question—or crisis—in the Mekong is an issue of scale. Basinwide, national and local interests are mutually constituting, and their interconnection is fundamental to an integrated water resource management framework. However, and contrary to the way in which scale issues are often considered in large scale basin management, it is not an issue that can be conceived of in terms of nested levels of administration and decision-making, or in terms of subsidiarity. Rather, the scale issue has to do with the paring apart of the notion of national interest to looking critically but also creatively at the implications of a transboundary governance framework in the Mekong that is based on multiple networks and diverse societal interests within and across borders.

On a final note, and despite the critical stance of this chapter and a report on which it draws,[1] the aim of this chapter is not to denigrate MRC. On the contrary, by devoting critical attention to the notion of national interest as an institutional underpinning and by thinking beyond such a uni-scalar level of cooperation and conflict analysis, the intention is to explore some of the current impasses that might be better understood, with an eye to more inclusive and multi-scalar transboundary governance.

Acknowledgments This chapter is based in part on research carried out with the support of the Australian Research Council and Danish Overseas Development Assistance. Please note that the original manuscript was produced in 2006.

References

Biswas A, Varis O, Tortajada C (2005) Integrated water resources management in South and South-East Asia. Oxford University Press, New Delhi

Blatter J, Ingram H (1998) States, markets and beyond: governance of transboundary water resources. Nat Resour J 40(2):439–473

Boesen J, Ravnborg HM (2003) From water wars to water riots: lessons from transboundary water management. Danish Institute for International Studies, Copenhagen

Ekbladh D (2002) "Mr TVA": grass-roots development, David Lilienthal, and the rise and fall of the Tennessee valley authority as a symbol for US overseas development, 1933–1973. Dipl Hist 26(3):335–374

Franks T (2004) Water governance—what is the consensus? The water consensus—identifying the gaps. University of Bradford, Bradford

He DM, Chen LH (2002) The impact of hydropower cascade development in the Lancang-Mekong Basin, Yunnan. Mekong Update Dialogue 5(3):2–4

Hirsch P (2001) Globalisation, regionalisation and local voices: the Asian development bank and rescaled politics of environment in the Mekong region. Singap J Trop Geogr 22(3):237–251

Hirsch P, Wyatt A (2004) Negotiating local livelihoods: scales of conflict in the Se San river basin. Asia Pac Viewp 45(1):51–68

Hirsch P, Jensen KM, Boer B (2006). National interests and transboundary water governance in the Mekong. Australian Mekong Resource Centre, School of Geosciences, University of Sydney in collaboration with Danida

Kliot N, Shmueli D, Shamir U (2001) Institutions for management of transboundary water resources: their nature, characteristics and shortcomings. Water Policy 3(3):229–255

McCormack G (2001) Water margins: competing paradigms in China. Criti Asian Stud 33(1):5–31

McCully P (1996) Silenced rivers: the ecology and politics of large dams. Zed Books, London

---

[1] In a recent study led by the author, the questions raised in this chapter have been addressed head-on in a policy context. The study is an applied policy study in the sense that it was developed and conducted between DANIDA, the largest donor to MRC since 1995, and the Australian Mekong Resource Centre (AMRC)—which has taken critical civil society concerns on board in its research on the Mekong. The full report and executive summary are available at http://www.mekong.es.usyd.edu.au/projects/mekong_water_governance2.htm.

Molle F, Wester P, Hirsch P (2006) River basin development and management. Water for food, livelihoods and environment: a comprehensive assessment. Comprehensive Assessment of Water Management in Agriculture, Colombo

Pante F (1996) Investing in regional development: the Asian development bank. In: Stensholt B (ed) Developing the Mekong subregion. Monash Asia Institute, Clayton, pp 16–21

Plinston D, He DM (1999) Water resources and hydropower. Policies and strategies for the sustainable development of the Lancang river basin. ADB TA-3139. PRC Asian Development Bank, Manila

Shah T, Roy AD, Qureshi AS et al (2003) Sustaining Asia's groundwater boom: an overview of issues and evidence. Nat Resour Forum 27(2):130–141

Shiva V (2002) Water wars: privatisation pollution and profit. Pluto Press, London

Svendsen M, Wester P, Molle F (2005) Managing river basins: an institutional perspective. In: Svendsen M (ed) Irrigation and river basin management: options for governance and institutions. CABI Publishing, Wallingford, pp 1–18

Thomas DE (2005) Developing watershed management organizations in pilot sub-basins of the Ping river basin. Office of natural resources and environmental policy and planning. Ministry of Natural Resources and Environment, Bangkok

Wolf A (1999) Criteria for equitable allocations: the heart of international water conflict. Nat Resour Forum 23(1):3–5

World Commission on Dams (2000) Dams and development: a new framework for decision making. Earthscan, London

Yang G (2004) Global environmentalism hits China. Yale Center for the Study of Globalization, New Haven

Chapter 4
Evolutionary Technologies in Knowledge-Based Management of Water Resources: Perspectives from South Asian Case Studies

Angela Barbanente, Dino Borri and Laura Grassini

Abstract Water management technologies and approaches in South Asia can be considered the result of two opposite forces: on the one hand, the modernist push for technological advance and reductionist thinking supported by local bureaucracies and international competition; on the other, local attempts to develop adaptive approaches and technologies, increasingly supported by NGOs and more recent international aid policies. Despite the disproportionate power between the two, evolutionary patterns of technologies are not the result of linear domination forces. On the contrary, local technologies and knowledge seem to evolve through complex interplay between local and global pressures as a result of cognitive and practical interactions. In the attempt to deal with these issues, our paper analyses some case studies from India. Our particular concern is with the evolutionary patterns of water management approaches and technologies with reference to the changing local–global interaction dynamics. In this context, we discuss the innovation potential of actual interaction spaces and the role local and global actor networks play in shaping mechanisms of cognitive interaction and technological innovation.

Keywords Local–global interplay · Technological evolution · Innovation · Cognitive interaction · Indigenous knowledge · Water management · Developing countries

The present chapter is the result of a joint research work carried out by the authors. Nevertheless, A. Barbanente wrote Sects. 4.1, 4.3, D. Borri wrote Sect. 4.6, L. Grassini wrote Sects. 4.2, 4.4, 4.5, 4.7.

A. Barbanente · D. Borri · L. Grassini (✉)
Department of Architecture and Urban Planning, Polytechnic of Bari, Bari, Italy
e-mail: laugrassini@libero.it

D. Higgitt (ed.), *Perspectives on Environmental Management and Technology in Asian River Basins*, SpringerBriefs in Geography, DOI: 10.1007/978-94-007-2330-6_4, © The Author(s) 2012

4.1 Introduction

International debate and aid policies on water technologies have been long dominated by the contraposition of two dominant paradigms, one pushing towards technological modernization and the other strongly resisting it through the development of alternative grassroot perspectives. The former dynamic is historically connected with colonization and with 'developed' nations, thus with unequal and unjust international relations. In particular, in India it refers to the diffusion of large dam projects, which have been considered to drive transformations of agriculture from subsistence to market, displacement of indigenous people, loss of cultural and social traditions, production of significant environmental damage (e.g. McCully 1996) and of predation of local ecologies and knowledge (Shiva 2001).

The latter dynamic is associated with the revival of traditional indigenous approaches to water management, which were developed by local populations before colonial times. Since these technologies and practices are considered to be in tune with local ecologies and social dynamics, their revival has become a corner stone in new global discourses on human and social development and sustainable resources management in poor countries (Parikh 1998; World Bank 1998). Despite great differences of these two paradigms, the boundary between the two is more blurred than expected. Strongly prompted by critiques of the technology transfer paradigm and by the aim to develop different approaches and technologies for water management, more recent development projects supported by international organizations entail small-scale schemes, which greatly rely on community involvement and decentralization in decision-making.

In this paper, after a critical description of the evolution of the international debate on water technologies in developing countries, a critique of the international approach to water management is made in Sect. 4.3. In particular, its centre-periphery models of innovation are discussed in relation to the capacity to grasp and favour real processes of innovation and change and effective learning mechanisms in a 'glocal' context. In Sect. 4.4 some examples of evolving technological systems within rural and urban Indian contexts are described, with a focus on the process of change and its dynamics. Their description is used to show how technological and cognitive innovations can be produced through a dynamic process of adaptation and change nurtured by the interplay of local and global actions; this is explained in Sect. 4.5. This analysis is taken a step further in Sect. 4.6 where attention is paid to the role of mediator-agents in the technical evolutionary processes as well as to the intertwining of cognitive, political and economic engines of transformation. Finally, some concluding remarks for challenging perspectives on technological evolutions in a 'glocal' context are reported.

4.2 Changing Technologies and Paradigms for Water Management in India: Local and Global Discourses

During the twentieth century, water discourses and practices in most developing countries have been largely influenced by the development of two opposite paradigms. On one hand, the transfer of technology paradigm was strongly supported by Western countries and by local bureaucracies, which aimed at strengthening their power as providers of primary services through a supply-driven approach (Gilbert 1994). In India, in particular, soon after Independence, the newly constituted country focused its attention on the development of large infrastructural projects, as a powerful means of eliminating the country's colonial past and driving it towards a socialist future of progress and prosperity.[1] Dams themselves became an icon of progress, being described as temples of modern India, and strong alliances between India and Western donor countries were drawn up (Hirshmann 1967; Black 1998). As a result, India is nowadays the third largest builder of dams in the world (Roy 2002). Moreover, the fast growing pace of resource exploitation associated with these projects, together with the growing conviction that it was the only way of gaining progress and success, increasingly put traditional small scale water harvesting technologies under serious threats.

While these dramatic changes were underway, another paradigm was slowly developing as a counterbalancing force. With growing international awareness of the disproportionate ecological and social costs of technological transfer paradigms based on standardized modern technologies (Escobar 1995; McCully 1996; Postel 1998; Guha 2000; Roy 2002), activists and researchers started searching for possible alternative solutions to water problems, based on the revival of traditional small scale technologies and practices of the poor, which had been used long before colonial times. Many studies mushroomed in the 1980s and 1990s, which depicted developing countries as repositories of alternative solutions, whose functioning in the past had guaranteed the survival of local populations even during situations of water scarcity and had moreover given them control over water resources and reduced their dependency on central schemes. The re-discovery of these systems was paralleled by the growing recognition of the importance of indigenous knowledge embedded in these technologies and representing a profound understanding of local ecologies (Chambers et al. 1989; Shiva and Bandyopadhyay 1990; Gupta 1998). This research, in fact, strongly opposed the assumption that non-Western reasoning was subjective and based on irrational

---

[1]   Soon after independence obtained in 1947, an average of 23% of public funds were given to the irrigation sector on a nationwide basis.

believes and myths[2] (Millar et al. 1999), and tried to demonstrate the intrinsic value of indigenous knowledge as a system of thought (Agrawal 1995).

In India there was a growing body of literature, devoting special attention to spectacular forms of water architectures of the past like stepwells, water tanks and other traditional systems (Mishra 1994; Agarwal et al. 1997; Sridhar 2001), as well as their potential for revival. At the same time, several small scale initiatives mushroomed at local level in support of alternative development projects, which tried to rehabilitate traditional technologies as viable solutions to current water scarcity problems. In particular, during the 1990s, an increasing number of NGOs were created, sponsoring these kinds of projects throughout the country in the name of the voiceless rural people. They then developed from isolated to networked phenomena. In the Gujarat state, for instance, a special network of NGOs working on drinking water issues was created at the beginning of 2000, as a pressure group lobbying for funds at the State level and a counterbalancing force against the dominant top-down State modernist planning projects.

A vivacious debate nurtured by modern and traditional paradigms for water technologies produced profound changes in international aid policies during the second half of the twentieth century. Theoretically speaking, the 'traditional' paradigm helped to relativize the widespread modernist rationality by suggesting that there were equally valid 'native' points of view. Politically, it contributed to the development of a great deal of literature on political ecology (Peet et al. 1996; Escobar 1996; Braun and Castree 1998), which challenged the assumption that the rural poor were somebody else's development strategy and the subjects, instead of being active originators of their own development patterns. Finally, on the technical side it supported the discussion on appropriate and low cost technologies against dominant approaches based on large scale infrastructural projects.

Since the late 1970s the term 'appropriate technology' seemed to synthesize the recipe of success for the creation of hybrid technologies combining modern inputs with indigenous practices. How far these were really able to answer local needs and foster sustainable solutions for water management is still unclear, as some authors underline the risk of constructing an alternative orthodoxy based on an idealistic picture of indigenous people and new standard techniques,[3] while showing a too simplistic faith in traditional technologies and an exaggerated critique of technological modernization (Bebbington 1996; Baviskar 1997). Nevertheless, these efforts in developing hybrid technologies continued in the

---

[2] These critiques developed because of the popular rhetoric on cognitive and epistemological differences between western and indigenous knowledge. Kloppenburg (1991), for instance, used paired concepts drawn from a range of sources to highlight differences between the two ways of knowing (e.g. tacit versus explicit, concrete versus abstract, intuitive versus rational, feminine versus masculine, craft versus science, relative versus absolute, indigenous versus scientific and so on).

[3] One example was the quick popularity of hand pumps in Indian rural villages, since they were considered an optimum compromise between simplified western technological input and local desire to manage water resources in a decentralized way (Black 1998).

1980s and 1990s (Grover 1998), when new collaborative bodies[4] were internationally developed in the attempt to produce learning spaces between local and global practices (WSSCC 1999). In this case, too, a key element for success was considered to be the definition of best practices, which increasingly became one of the main cornerstones for the development of aid policies in developing countries (Emmerij 2002; World Bank 1998, 2001).

Within this changing global context, where different paradigms are being confronted and where international discourses and practices seem to be oriented by the definition of good and bad practices as guiding principles, how far are they able to foster innovations and sustainable solutions for water management?

4.3 Beyond Centre-Periphery Models of Innovation Diffusion in Water Management

In this section, we outline the essential theoretical and conceptual framework that we use for analyzing questions of innovation diffusion in water management in the following case studies from India. The starting point is the awareness of inadequacy both of the well-established centre-periphery model for the diffusion of innovation in the 'glocal' economic and cognitive geographies that characterize contemporary urban-regional spaces (Swyngedouw 1997) and the deterministic perspective on technology, according to which technology is an autonomous development force that dominates society (Severino 1998).

The centre-periphery model has been widely used not only as an explanatory model to account for social differentiation and technological change at the global level, but also as a normative model to promote diffusion of innovation in developing countries. This model is at the core of water management technologies and approaches underlying the transfer of technology paradigm. It is based on three essential elements: (i) the innovation to be diffused exists prior to its diffusion; (ii) diffusion is the movement of an innovation from a centre out to its ultimate users; (iii) directed diffusion is a centrally managed process of dissemination, training and provision of resources and incentives (Schön 1973, p. 81).

The centre-periphery model has been considered simplistic from different viewpoints in the light of the complex knowledge relations in contemporary society. From a phenomenological perspective, Schön noted that the essence of the core/periphery model fails to capture the complexity of system-wide social change in which innovations originate from numerous sources and evolve as they are diffused. On an anthropological base, the notion of a cultural centre dominating

---

[4] Among them, two of the most important are the Global Water Partnership and the Water Supply and Sanitation Collaborative Council, both promoting global forums and smaller scale working groups with the aim to foster a dialogue among governmental and non governmental bodies, donor organizations, professional and research bodies, and the private sector.

the periphery has been questioned too. Two reasons are worth mentioning here: firstly the possibility to single out a 'fixed' core is inherently limited in an age characterized by globalized interrelationships, and secondly the implicit postulation that the periphery is unable to resist the cultural domination of the centre. Appadurai (1996) highlighted the inherently simplistic scope of the centre-periphery model by emphasizing the multiple non-economic flows between individuals, groups and states. He focused on a multitude of exchanges and agents in the economic, political and cultural realms, and analyzed the processes of globalization with respect to the impact that they have on the individual and other micro agents both at local and global levels.

Here the concept of 'glocal' comes into play. This conceptual move was largely based upon the observation of empirical practice. The term hints at the relation between globalism and localism in contemporary society. It expresses both the need to go beyond the line of thinking that argued in favour of globalization as homogenization and the need to grasp the possible coexistence of a global homogenized culture and local cultural traits. The deriving concept of 'glocal-ization' (Swyngedouw 1992) indicates the combined process of globalization and local-territorial reconfiguration, with an emphasis upon the changing relationships between local and global. It refers to (i) the contested restructuring of the institutional level from the national scale both upwards to supra-national or global scales and downwards to the scale of the individual body, the local, urban or regional configurations and (ii) the strategies of global localization of key forms of industrial, service and financial capital.

The concept of glocalization rejects the assumption that globalizing pressures are in conflict with local cultural identities as it is arduous to conceptualize the global as if it usually and unavoidably excludes the local. The local is embedded within, and superimposed upon, the global, while global processes seem to pervade all expressions of the local (Amin and Thrift 1994). From a cultural anthropological point of view, the concept of glocalization emphasizes processes of global creation of the local and localization of the global. It is the intersection of such directions of flows that brings together the global and the local giving rise to the formation of new hybrid cultures.

More generally, we face a hybridization process when local traditions become transformed by the addition of ideas or elements borrowed from elsewhere. The notion of 'hybridity' is clearly connected to the ideology of postnationalism. If nationalism has all the connotations of fixity and repression, then hybridity, according to Bhabha (1994), captures the liberation potential of resistant cultures. Hybridity counters the dominant logic of authoritarian discourse and opens up the third space, the interstices where meaning is always in-between never stable, never rigid. As Ritzer (2003) notes the notion of glocalization bears clear parallels to Appadurai's (1996) concept of hybridity.

The concepts introduced above induce us, empirically, to consider the economic and cognitive flows that interconnect local and global spaces, which are much more intricate than the linear and unidirectional ones postulated in the centre-periphery model. As far as the evolution of water management approaches and technologies is concerned, these economic and cognitive flows affect all the agencies involved: powers of globalization and market competition, supra-national institutions such as the World Bank, national government institutions operating at different levels, cross-border cooperative organizations and local government institutions, specialized authorities carrying out specific functions, private business enterprises, and local community groups.

As a consequence, water supply and sanitation projects emphasizing community involvement and decentralization cannot be considered exclusively local and immune from influences coming from the centre. They have to be analyzed in the light of the penetration of such notions in supra-national institutions and cross-border cooperative organizations that were usually confined to the sphere of global thinking. This implies that water supply and sanitation technologies based on traditional local practices may be 'globalized' through the diffusion of 'best practices'. On the other hand, we should take into consideration the dynamically conservative plenum into which information moves from who knows about the innovation to who is supposed to get it and ask ourselves how the ponderous bureaucracies of South Asian water management institutions face problems created by new situations.

A local culture can never be kept in its original form. It should be understood as a dynamically constituted outside practice. Knowledge is a process of 'enskillment' in the context of practical engagement with the environment, and local knowledge is "a practical, situated activity, constituted by a past, but changing, history of practices", a set of time and context-specific improvised capacities rather than a coherent "indigenous knowledge system" (Escobar 1998, p. 62, quoting Hobart 1993, p. 17). Social systems are based on knowledge that can be both formal, codified (scientific or engineering knowledge) and informal, tacit (embodied in skilled personal routines or technical practices), and the latter is very important in explaining skillful performances of human beings (Argyris et al. 1985, p. 49). The empirical investigation summarized in the following section focuses on changes that happen in practical, situated processes of water management.

4.4 Local Technologies in Evolution: Some examples from India

In order to gain further insights on the issue of technological innovation in water management, a micro-level perspective is adopted in this section, through the analysis of some case studies of technological evolutions within rural and urban Indian contexts. They have been analysed through local interviews with a large number of stakeholders, ranging from local villagers to representatives of NGOs and of local authorities in charge of water supply.[5] Their analysis is aimed at

---

[5] If specific sources of data are not cited, information contained in this section is directly taken from field observations and interviews.

pointing out innovation patterns within the technological evolutions observed, as well as the factors really influencing these within changing global–local dynamics.

The first example comes from one of the most arid areas of western India, the Saurashtra region, in the Gujarat State. Due to scarcity of surface water, in that region water is traditionally harvested through dug wells. However, the performance of this technology reduced during the 1970s and 1980s because of high groundwater extraction and low soil permeability. This resulted in significant losses of harvests throughout the region. Driven by this situation, people started thinking of possible ways to increase groundwater recharge by channelling monsoon runoffs and diverting it into dug wells, in order to increase groundwater level. For this purpose, they started constructing small check dams, in association with some diversion schemes. Check dams themselves were not new in that region, since they had been traditionally used by local people to harvest runoff into seasonal reservoirs for agricultural purposes. But in the experiments of the late 1970s that system was used in a different way from the past. Through multiple trials, villagers realized that if water was not left in the reservoir but channelled into lower soil strata it could effectively produce groundwater recharge and increase groundwater availability for the following years. Because of this, they decided not to harvest water directly from the reservoir, as they had done in the past, but to divert it under the ground and then use wells to extract it. Personal communication and some awareness campaigns were organized by different grassroots movements[6]; then these technologies and underlying knowledge systems spread through the region (Shah 2000; Prasad 2001).

Due to this raising of awareness of the positive impact of the use of these water management systems, the Gujarat State even launched a special funding programme in 1999—the Sardar Patel Participatory Water Project—with the aim of funding the construction of 2100 check dams in the region. The availability of additional funds from the National Bank for Agricultural and Rural Development then led to a strong increase in the final number of funded projects, so that, by the end of the year 2002, 13,000 check dams had been built instead of the planned 2100 (GEC 2002), so making their final number in Saurashtra more than seven times higher than it was in 1999.[7] This probably produced great impacts on the local ecosystems, although a detailed analysis was never carried out by either the State or by the several NGOs actively engaged in the process as implementing organizations. They all seemed much more interested in following a 'best practice' approach and channelling as many funds as possible.

In this context of growing attention (and funding) for these experiments, some Research Institutes, too, became interested in recharge issues in Saurashtra. Among these, the Indian Space Organization launched a programme for satellite

---

[6] Grassroots movements range from religious groups, acting on the basis of spiritual belief in conservation practices, to community based associations later developed into the Saurashtra Lok Manch NGO.

[7] In 1999, the total number of check dams in Saurashtra was around 2000.

image interpretation aimed at finding the most effective points of groundwater recharge, mainly coinciding with intersections of fissured lines. These pieces of information was then used by some villages to further improve the effectiveness of their recharge technologies within the catchment area.

As a result of these multiple efforts, ranging from self-made experiments of local people, to grassroots and spiritual drives, to National and State funding schemes, to scientific research programmes, something interesting was produced in the region. Many villages started looking at these initiatives as a source of unexpected and fruitful resources. These programmes and actions, whatever their initial aim and rationale had been, were channelling important things to that forgotten land: funding, technologies and knowledge. It was up to the local people to find a way to make the best out of this situation in order to produce new solutions to their problems. They then started combining selective pieces of technologies and knowledge from different sources in a dynamic local context, thus producing technological systems which were neither truly traditional nor modern, but a mix. In this way they also showed their creative and innovation potential.

Another example of this proactive role of local people in the definition of evolutionary patterns of technologies and knowledge systems for water management can be found in the analysis of a slum upgrading process in Ahmedabad city. Slums are usually the most deprived parts of cities mainly encroaching in depressed areas and lacking even the most basic infrastructures and public services, also because of their illegal status. Water availability is a very problematic issue. In these contexts, not only do people lack connections to urban pipelines but they cannot even rely on traditional water harvesting technologies, which seem inapplicable due to overpopulation. Because of this, water resources mainly come from a few illegal connections to municipal schemes, around which many conflicts arise within the community.

Despite increasing attention paid by current theories and practices of slum upgradation to participatory mechanisms and community-based solutions,[8] the need to provide modern—even if low cost—technologies and connections to urban infrastructures is generally supported. What are the implications of the introduction of new technologies? While it might be seen as an attempt to introduce key stereotyped solutions based on modern technologies, the results of these projects may not necessarily entail the adaptation of local practices to external rules as local people may give unexpected input to transformation and technological change, as will be shown in this case study.

In Ahmedabad, the slum upgrading project followed the Slum Networking approach, which was developed by an Indian engineer, Himanshu Parikh, and

---

[8] While in the 1950s and 1960s the 'slum clearance' approach prevailed and led to great demolition processes and relocation strategies in developing countries, in the 1970 s more comprehensive upgrading strategies emerged, where house rehabilitation was part of a broader process of community development (Cohen 1983). More focused sectoral strategies then emerged around the Nineties, with increasing attention being paid to institutional and participatory mechanisms (Kessides 1997).

already implemented, with minor modifications, in other two Indian cities between the end of the 1980s and the end of the 1990s[9]; this approach was also receiving increasing international attention and recognition as global best practice[10] (Grassini 2007).

From a technical perspective, it was based on the idea to use the improved infrastructural connection of slums as an opportunity for a quantum change in the network functioning of the city as a whole.[11] This core technical idea was then supported by a partnership approach among the slum community, a local municipality and a third party, as a way to ensure resource self-sufficiency and community control in the planning and management phases of the project.[12] The observations on which this paper is based are taken from Sanjay Nagar, a slum colony in Ahmedabad where the project was piloted before its scale-up at city level.

Beyond the international rhetoric of a best practice initiative, what is more interesting for our purpose is a close analysis of the way technologies were actually implemented and used in practice in this slum. While the project effectively led to the provision, to all dwellers, of individual water connections and toilets, concrete streets and street lighting, the final features and use of these technologies were innovated by local people after project completion. Individual toilets were not only used for sanitation purposes, but also as shower closets, thanks to the use of baskets of water taken from the nearby water connection. The concrete streets, although paved and straightened to channel away rain water and to allow easy placement of underground sewerage networks, became much more than this. Slum dwellers covered them with branches and transformed them into 'verandas', which became extensions of the houses and socialization places to cook or rest during the hot summer days and where children could play out of the dust safely. Finally, thanks to the opportunities offered by this project, traditional

---

[9] The Slum Networking Project was implemented in 183 slums of Indore between 1989 and 1997 and in some pilot schemes in Baroda in 1994. Implementation in Ahmedabad started in 1996 (Tripathi 1998).

[10] The Slum Networking Project in Ahmedabad, Indore and Baroda received several international awards, like the Best Practices award at the Habitat II conference in Istanbul in 1996 and the Aga Khan Award for Architecture in 1998. It was among the best practices list of Habitat II in 1998 and part of the best practices examples in the Cities Alliance, as a global partnership among cities launched by the WB-UNCHS in 1999 with the final aim to be freed from slums and to make a contribution to the improvement of the living conditions of 100 million slum dwellers by 2020. For further details see Grassini (2007).

[11] Since slum areas are usually illegally built along main drainage patterns of the cities, the improvements of their drainage and sewerage patterns and their connection to the city main network should result in a general improvement in the gravity based infrastructural network of the city as a whole (Parikh 1995). From here the project name, 'Slum Networking'.

[12] In Ahmedabad, the partnership envisaged three major groups: the Ahmedabad Municipal Corporation (AMC), the local community and a private industry, with additional support from a local NGO and Parikh himself.

communal practices, previously abandoned in the degraded slum environment, could be recovered in different forms and brought to new life.

The introduction of new technologies from outside thus nurtured a more profound process of technological change, where old and new practices and technologies were re-interpreted by local people thanks to their ability to turn traditional and external inputs into new ones. In this way they proved to be far from passive recipients of external aid and stereotyped technologies.

4.5 *Bricolage* as a Mode of Innovation and Change: Cognitive and Practical Implications

While the international debate is mainly focused on the contraposition between global scenarios of development and indigenous alternatives, the cases from India described above show how interesting results may come from a sort of practical interaction between local and global inputs, where pieces of technologies and practices join both indigenous and western traditions. This resembles the practices long observed by anthropologists, where local people could effectively transform rules superimposed by conquerors, through a sort of *bricolage* strategy and piecemeal approach of adaptation and change (de Certeau 1990; Appadurai 1996). This also points to the concept of 'cognitive *bricolage*' as developed in the studies on organizations, where the selective use of routine and practices within new contexts has been observed (Lanzara 1993).

In this perspective, the case studies suggest that modernity and local traditions are not black boxes which collide, but they can be deconstructed and reshaped in everyday practices of poor people, who are able to use even insignificant pieces and materials to invent new and hybrid means for their survival. The analysis of this kind of practice can give helpful insights for the interpretation of technological evolutions in the water field.

First of all, unlike international debate on the appropriateness of traditional versus modern technologies, these practices suggest that innovations and changes in local patterns of technological use are not strictly pre-determined within implemented technologies. On the contrary, local people can produce new evolutionary patterns of technological change through which modern and traditional pieces are reshaped and combined into new hybrids. In this context, what is also interesting is that the initial input for transformation can come both from a traditional technology, as in the Saurashtra case, and from a modern one, as in the case of the slum upgrading programme.

This integration itself can be considered the essence of what some authors call "regional modernity" (Sivaramakrishnan et al. 2003), as a concept "where localities are the 'places', both real and figurative (village landscapes and people's minds) where modern knowledge systems are continually produced using information gleaned from local conditions, drawn through the lens of a larger, regionally-based culture and historically-determined circumstances,

and broadened with information gained from globally circulated sources."
(Brodt 2003, pp. 344–345). In this context, locality is not a unitary, homogenous
entity founded in some perceived idea of indigeneity to which it is possible to
ascribe some specific practices and technologies. It is, indeed, a contingent entity,
historically situated and related to specific dynamics which are not all confined
within the locality itself. Local people themselves are far from static, but curious
and willing to pick up ideas from different sources and even to innovate by
accidental discoveries (Critchley 2000).

Innovation itself is produced by a silent and slow *bricolage* activity of adaptation and change put forward by local people in such a way that what is actually
produced cannot be considered either indigenous or modern, but something
different (Grassini 2002). In so doing, local people prove to be able to think across
different contexts, i.e. to apply routines, practices and technologies to contexts and
purposes different from those traditionally associated with them. They thus seem
to use what is usually called 'improvisational capability', i.e. the capacity to foster
change in adaptive and creative ways in between external and internal pressures.
According to this perspective, innovations are the result of neither deliberative nor
adaptive approaches, but they often stem from the capacity to innovate and
change, which lies at the intersection between the two (Weick 1995, 2001;
Brunsson et al. 1998).

This broader understanding of innovation as a social, nonlinear and interactive learning process implies a change in the evaluation of the role played by
local socio-cultural structures in technological development, from being looked
upon as mere memories from pre-capitalist societies, to be considered as
necessary fundamentals for innovation in a post-Fordist learning economy
(Asheim 1999, p. 347). When dealing with water management approaches and
technologies in increasingly 'glocal' contexts, we should not consider local
solutions as isolated, autonomous localities. The need is for differentiated,
responsive, continually changing, but connected reactions (Schön 1973,
pp. 188–189). In this vein, insisting on the promotion of 'local solutions', with
an emphasis on radical decentralism as a response to the forced uniformity and
perceived distortions of centre-periphery systems, can create as many problems
as they solve.

The acknowledgement of these processes of change, thus, poses further issues
in relation to the cognitive implications of such changes. While technologies are
reshaped in practice, how does local knowledge about them evolve? How are
hybrids of knowledge and technologies co-constructed from indigenous and
external inputs? In order to interpret our Indian case studies, we refer to recent
research in the anthropological field, according to which what happens during
these evolutionary processes is a slow process of hybridization of the different
hierarchical levels of abstraction that compose the knowledge system. In particular, they demonstrate how local hybridization processes between knowledge
of different geographical and socio-political origins can either bring new
primary-level information from an outside source to be subsumed under existing
secondary-level interpretative concepts; or they can lead to the introduction of new

secondary-level concepts, which expand or explain the existing system[13] (Brodt 2003).

In the Saurashtra case, for example, new practices—like the diversion schemes within dug wells—and their cause-effect relationships have been imported within the traditional interpretative framework of recharge/withdrawal equilibrium. Instead, lately acquired secondary-level concepts—like the effectiveness of recharge in relation to the fracturation ratio—were adopted to expand the potential of existing recharge practices. How these pieces of knowledge will be combined in a new evolving knowledge system will then also depend on different factors which influence the usability of information at different hierarchical levels.[14]

4.6 An Agent-Based Perspective on Technological Change

The inadequacy of binary—hard/soft, traditional/modernist or post-modernist—descriptions of technological models of exploitation of water resources in our case studies from India contributes to raising further theory-in-practice issues. In particular, the intriguing evolution of water technologies as they emerge from case studies and the frequent appearance of hybrid solutions and processes have prompted us to focus attention on these relationships and on the agents involved.

In fact, in tune with theoretical and empirical literature, our case studies show that (i) in most communities water technologies are not naively derived from purely local tradition but evolve according to a context interaction in order to better tune community technological behaviours towards endogenous and exogenous needs, as in any other evolutionary process of adaptation, to the dynamics of the environment, and (ii) hybridization of traditional and modern-standard technologies in this evolution is a process that often leads to innovative techniques in intriguing non-linear terms which deserve better consideration (Norton and Bass 1992; Dattée and Birdseye Weil 2007).

The process of technological change (TC) oscillates between transition (a radical innovation and in-depth movement in technology involving multiple complex systems) and incremental change (almost imperceptible movement deriving from small changes often consisting of adapting existing technologies to

---

[13] By primary level we mean the level consisting of pieces of information about the objects and their cause-and-effect, spatial, and temporal relationships; by secondary level we mean the level consisting of the overlying 'concepts', which are abstracted from the information level and explain the information pieces in a meaningful whole.

[14] In this respect, some studies (Brodt 2001) have demonstrated that a knowledge system as a whole may survive better in the face of changes if it is organized into relatively independent subsections which can have their own independent pattern of development. On the other hand, the erosion of traditional knowledge is also related to the scale of application, the possibility being to scale-up practices which are likely to be maintained even within changing economic circumstances. This is particularly relevant given the unequal scale power that modern technologies can have because of their mode of production and ease of diffusion.

new contexts). The concept of TC transition and of technological trajectory (Dosi 1982, cited in Elzen and Wieczorek 2005, p. 653), in particular, has recently become object of widespread research interests because of its appropriateness to situations where radical technological changes are occurring or are needed in terms of sustainability processes.

TC transition systems (or 'regimes', as some scholars prefer to define them to stress social implications vis-à-vis the technical ones better) are based on multiplicity and co-evolution of actors, factors, and levels. Traditional interpretive policy-oriented paradigms of TC consider a tripartite model: top-down (authority based, political), bottom-up (market based, economic), networked (knowledge based, cognitive).[15]

Reflections on our Indian case studies confirm what recent complexity system approaches to TC maintain, that in the real world various dimensions of the technical change overlap—cognition can derive from a political effort of innovation as much as economic valuations can be influenced by political incentives or disincentives like in the Saurashtra case, or conversely, political push for technical innovation can derive from the evidence of economic advantages of some technical innovations as in the slum case—so that none of the individual dimensions of the above mentioned hyperspace of technical change is completely meaningful for the understanding of the evolutions of techniques. The agent-based perspective— focusing on both the formation of elements and paradigms and the relationships among elements and paradigms in the technical space—presents an evolving knowledge-in-action and global–local situation of techniques which appears adequate to the integral understanding of the phenomena at hand (Dattée and Birdseye Weil 2007).

Our agent-based perspective tries to overcome another major critique to the analysis of TC transition systems-regimes, which is the almost exclusive focus on producers and the ignorance of users' roles and judgments (Elzen and Wieczorek 2005). Very little literature is devoted to common sense knowledge generation and evaluation on TC and it often reports in too aggregate forms about actions—more

---

[15] Simplified modelling of technological change is of course trivial, in the light of the high complexity of situations and dynamics: it can help to understand the structure and functioning of the technical space under consideration. New techniques replace the old ones according to a process which evolves in a multidimensional space. When this hyperspace is reduced to a 3D— cognitive, political, and economic—space, three mechanisms (macro-agents) rule this evolutionary technical space: (i) cognition of a new technique somewhere replacing an old one is precondition for the adoption of a new technique elsewhere (or in given multi-agent space) and for building organizational environments around it; (ii) politics—endogenously or exogenously driven—influences penetration of a new technique into the space of an old one; (iii) economy allows the replacement of a given technique with a new one on the basis of the economic convenience or inconvenience of this replacement. This schematic model is, partly, an agent-based transposition of the well known technology life cycle model (Moore 2002; Rogers 2003). At the top level of the technological arena macro-agents (institutional, social) continuously operate technological change within this technological multidimensional space while at the bottom micro-agents (technique raiders-representatives, individuals) continuously implement and contribute to the evolution of this technical change (Henderson and Clark 1990).

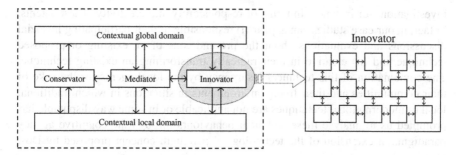

Fig. 4.1 Conservator-mediator-innovator technological agents

than reflections—from below for pushing or pulling or opposing TC. Even the intriguing concept of tribal technology (Shiva 2002) seems still to lack empirical robustness to be able to use it as a sound basis for further investigation on dilemmas of judgments on TC.

Moreover, agent-based cognitive approaches are spreading generally in the analysis and modelling of organizational development and planning processes, in particular when processes of social learning and technological change are involved. These approaches (O'Hare and Jennings 1996; Ferber 1999; Forester 1999; Borri et al. 2008) being oriented to exploring the behaviour of both the single-agent and the multi-agent (of a cognitive agency at different scales of amplitude of the group of agents involved), they fit particularly well to situations in which organizational and community development continuously stems from a complex set of individual and social cognitions and actions (Fig. 4.1). Well formalized economic research, too, is focusing on the cognitive role and credibility of agents—in particular experts with different skills and information—in decision-making in multi-expert environments (Feinberg and Stewart 2008).

The individual-mediator-community distributed agent basic model (Okamoto et al. 2009) seems to fit particularly well to the analysis of technical changes in which two different actors—the rational individual agent of the expert standard techniques (powerful because of its oversimplification), on the one hand, and the emotional social agent of the commonsense techniques, on the other, confront each other by the interposition of a mediator-agent in between them, an agent which masters the typical problems of coordination and conflict resolution that arise in the theory-in-practice dialogue between One and Two (Shakun 1999). Further complexity comes from the recognition that individual expert rational agents are indeed encapsulated in their own social network while groups of commonsense emotional agents can appear as fragmented in their own constituency of individuals, with the obvious consequence that 'individual' and 'social' are intriguing intertwining cognitive categories.

First of all, challenging the idea of techniques as Leviathan agents which develop and function out of human control (Severino 1998), our case studies show that local agents are not overwhelmed by techniques as passive objects but have interesting proactive roles in technological change which requires further investigation, for instance, in terms of responsibility and creativity. These agents, in fact, in our case studies can appear (i) responsible in their participating in social discussions and evaluations about the pro-and-cons of an existing or proposed technique and (ii) creative in their radically transforming an existing technique. The combination of this technical insurgency by the local agents for the sake of their communities and their living environments in situations in which traditional local or modern global techniques are not applicable or in some way disputable has prompted us to analyse these technical behaviours through the cognitive agency paradigm, an extension of the technological paradigm concept proposed by Dosi (1982).

Transition Management (a learning-by-doing process in which "different participating actors from society work on inducing a specific systems change") (Kemp and Rotmans 2004, cited in van de Kerhof and Wieczorek 2005, p. 734) implies organizing transition arenas for developing visioning and learning in their various contexts—social, political, government, organizational, etc.—and hierarchy and abstraction levels—first-order, dealing with options that do not require radical changes in theories and/or practices, second-order—and many other conceptual and practical aporias (van de Kerhof and Wieczorek 2005).

If structure and evolution of technical systems are conceived from an agent perspective, agents appear involved in transactions which imply a relational understanding and change of the alleged basic elements and paradigms of the technical systems at hand. Located at the peripheries of these, in intermediate spaces of the system spaces, and garrisoning cognitive and operational relationships among elements and paradigms elaborated by other organizations, these agents do not uniquely participate nor are they restricted in any of the individual technical elements and paradigms on which given domains of techniques are built. On the contrary, potentially they are active transformers and mediators (for instance between top down technological agents belonging to institutional technical organizations and bottom up technological agents belonging to communities and grassroots organizations) of relationships and evolutions among different technical elements and paradigms—that is among different agent domains—and contribute to build 'glocal' knowledge about those technical domains (Pendharkar 2007). They continuously execute typical mediator tasks, like understanding positions of different agents in a technical arena, introducing their own interest into the arena, and also transforming the arena according to their mission of coordination (Okamoto et al. 2009).

Any set of users of an old technique is differentiated on the terrain of the users' various degrees of conservation—or innovation—propensity, while those conservation/innovation (behavioural) ratios that are expressed by agents and contexts define the pace of replacement of the old techniques by the new ones available.[16]

---

[16] In this perspective, religious and other symbolic attitudes towards science and technology could be considered as greatly influencing the alternation of scientific and technological visions and arrangements.

Nonetheless, as mentioned above, in the last few years it has become increasingly clear that many processes and events[17] can bring back the reuse of old techniques, obviously in ways that fit contexts often profoundly different from the past, thus denying visions of unidirectional evolutions of techniques. In this circular stream of technical progress, the reuse of old techniques can mean selecting technical primitives—as technical ontologies—that are fundamentals and invariants in the artificial (i.e. implying intentional transformation) relationship between the environment and its human beings (Antonelli 2007).

While the evolution of techniques in this conceptual hyperspace appears to us as ruled by either informed decision about feasible alternatives (active change) or influence of contextual events of selective decline-and-rise in domains of techniques (passive change), in the real world things are fuzzier and less discontinuous. Replacements of techniques mainly function in a market-like space of techniques, so that there is usually coexistence of old and new techniques within one given socio-geographical context as well as within a plurality of socio-geographical contexts. It is worth noting, this structural inertia of any technical change also leads to a cognitive vantage, given the possibility of reflecting on the technological evolution in relation to a given problem and of introducing further dynamism in the process.

Nevertheless, the coexistence and alternation of traditional and innovative techniques and in general of different techniques in the course of the evolutionary multi-agent technical process pose problems of management and co-ordination: as in any multi-agent process, the role of mediators-agents becomes essential for managing and co-ordinating the cognitive, political, and economic implications of the evolutionary process; in fact, mediators-agents are actively located in between the two original groups of senders and receivers of technical change transactions and work for continuous re-creation of technical frames. Mediators-agents operate important proactive roles of technical communication, evaluation, and assistance. They stay in between the traditional techniques and the innovative ones and contribute to the cognitive framing of the transaction space of the evolutionary technical process.[18]

Mediation and transaction of techniques continuously build an articulated technical space in which knowledges and organizations shape economic, political, and environmental interpretations of advantages and disadvantages of technical situations. Moreover, this supports deep reflections within the ecological debate in the globalizing era, where discussion between traditional and innovative techniques highlights important concerns. In this context, in fact, while many contributions are focusing on the critique of the effects of technical innovations on

---

[17] See the rise and the development of the ecological movement, which has highlighted failures of, and ecological risks from, new techniques (Worster 1985).

[18] The acknowledgement of their substantial role does not mean ignoring the fact that relevant doubts have been raised about the ability of agents to control the evolution of techniques in their compulsory and dominating relation with the world (Vattimo 1997; Severino 1998; Galimberti 1999).

the environment with a tendency to technical conservative attitudes,[19] a growing body of research is now trying to combine the attention for the preservation of nature with that for organizational and technological changes in view of increasing efficacy and efficiency (Ketelar and Todd 2001; Gifford 2005).

4.7 Concluding Remarks: Challenging Perspectives on Technological Evolutions in 'Glocal' Contexts

In this paper we have tried to show how technologies may undergo a slow evolutionary process of change as a result of active interactions between communities, techniques and their underlying knowledges and practices. For this purpose we have selected two cases from India, whose main interest lies in the process which they point to and not in the specific technological output they produce. Our aim was to contribute to a rethinking of planning interventions in South Asian countries, which are often polarized between two opposed normative approaches, the modernist and the traditional. Although it is almost impossible to derive recommendations for such a wide and diverse area, some interesting considerations can nevertheless be drawn.

First of all, in contrast with the dominant idea that technologies drive communities towards a pre-defined pattern of change, which they can at the most try to resist, this paper demonstrates that technologies themselves may be only the first input in a local process of innovation and change. Through this process, which often goes unnoticed, technologies themselves are deconstructed and reinterpreted in innovative ways, thus leading to new patterns of change which are nurtured by the dynamic relationship between local and global pieces of technologies, knowledge and culture.

The acknowledgement of these innovation potentials 'in the making' of the process throws a different light on current attempts of many international organizations to define hybrid technologies as a result of a mix between efficiency-led modern inputs and contextual-based indigenous practices. While these attempts are driven by the idea that some sort of definition of best practices is possible and that these practices will be able to spread innovation potential according to well defined patterns, we have tried to demonstrate how technological inputs are often only one of the manifold starting points of a slow process of change, where local and global actors and inputs actively engage in transformation through a dynamic relationship. In this perspective, any real attempt to foster planning processes that wish to rely on grassroots potential for the definition of new patterns of change needs to acknowledge the importance of local people's agency instead of any stereotyped output of that agency. This means that an idealistic defence of specific indigenous technologies is not the right way to go. Instead, we must work for the

[19] In this view, globalization is mainly seen as bringing homologation of techniques and therefore decontextualization and deterritorialization, i.e. dangerous detachment from the structures and constraints of the local environments.

enhancement of local people's possibility to actively engage in transformations through negotiations and context-based interpretations of local and global inputs.

This has very important consequences. First of all, it implies the acknowledgement of the pro-active role of local people in the definition of evolutionary patterns of technological change. Of course we do not mean that local people agency needs to overcome all others—this is also inconsistent with the dialectical relationship between local and global forces and inputs. In our contribution, we speak of the need for negotiations between local and global agencies instead of the need to have local people dominate the choice. And this is also why we stress the importance of mediators-agents, who operate important proactive roles of technical communication, evaluation, and assistance in between local–global transaction in a complex knowledge/technological space.

In this perspective, border zones prove to be the most favourable areas for transformation, where multiple inputs are available for interaction and hybridization processes, and where these can enrich the range of possibilities for future actions. Out of the dualistic contraposition of domination and resistance between traditional and modern technologies, innovations seem to come from a sort of "border thinking" (Mignolo 2000), as a thinking happening in places where diversities can meet and hybridizations can happen in everyday practices through piecemeal and silent processes of change instead than of grandiose and deliberate actions. Actual 'glocal' spaces may, thus, become the crucial areas for innovations, beyond their traditional interpretation as places of domination and resistance between local and global forces. On the contrary, they can be seen as places where fruitful processes of evolution and change can take place, nurtured by local and global inputs (Geschiere and Meyer 1998; Escobar 2001; Borri et al. 2004; Barbanente et al. 2007). In this context, Asian societies, which often look as a composite repository of traditional and modern technological systems, can find an important source of vitality and strengths for future innovations and changes.

Acknowledgments The empirical material on which the paper is based has been developed as part of the PhD research of Laura Grassini in Urban and Regional Planning at the University La Sapienza, Rome, with funding from the same University. A first draft of this paper was presented at a special session on "Changing Perspectives on Technology in Environmental Transformation and Management" of the conference "Asian Horizons: Cities, States and Societies", held in Singapore with funding support from the National University of Singapore. Special thanks are due to David Higgitt for the organization of that session and the opportunity for in-depth discussion.

References

Agarwal A, Narain S (eds) (1997) Dying wisdom. Rise, fall and potential of India's traditional water harvesting systems, State of India's environment: a citizens' report, 4, Centre for science and environment (CSE), New Delhi

Agrawal A (1995) Dismantling the divide between indigenous and scientific knowledge. Dev Change 26:413–439

Amin A, Thrift N (1994) Living in the Global. In: Amin A, Thrift N (eds) Globalization, institutions, and regional development in Europe. Oxford University Press, Oxford, pp 1–22

Antonelli C (2007) The system dynamics of collective knowledge: from gradualism and saltationism to punctuated change. J Econ Behav Organ 62:215–236

Appadurai A (1996) Modernity at large: cultural dimensions of globalization. University of Minnesota Press, Minneapolis [Italian translation: modernità in polvere. dimensioni culturali della globalizzazione. Meltemi, Roma, 2001]

Argyris C, Putnam R, McLain Smith D (1985) Action science: concepts, methods, and skills for research and intervention. Jossey-Bass, San Francisco

Asheim BT (1999) Interactive learning and localised knowledge in globalising learning economies. GeoJournal 49:345–352

Barbanente A, Camarda D, Grassini L, Khakee A (2007) Visioning the regional future: globalisation and regional transformation of Rabat/Casablanca. Technol Forecast Soc Chang 74(6):763–778

Baviskar A (1997) In the belly of the river. Tribal conflicts over development in the Narmada valley. Oxford University Press, New Delhi

Bebbington A (1996) Movements, modernizations and markets. Indigenous organizations and agrarian strategies in Ecuador. In: Peet R, Watts M (eds) Liberation ecologies. Environment, development social movements. Routledge, London, pp 86–109

Bhabha H (1994) The location of culture. Routledge, New York

Black M (1998) Learning what works. A 20 year retrospective view on international water and sanitation cooperation, report, UNDP-WB water and sanitation program

Borri D, Camarda D, Grassini L, Barbanente A, Khakee A (eds) (2004) Local resistance to global pressure. A Mediterranean social/environmental planning perspective. L'Harmattan, Paris

Borri D, Camarda D, De Liddo A (2008) Multi-agent environmental planning: A forum-based case-study in Italy. Plan Pract Res 23(2):211–228

Braun B, Castree N (1998) The construction of nature and the nature of construction: analytical and political tools for building survivable futures. In: Braun B, Castree N (eds) Remaking reality. Nature at the millennium. Routledge, London, pp 3–42

Brodt SB (2001) A systems perspective on the conservation and erosion of indigenous agricultural knowledge in Central India. Hum Ecol 29(1):99–120

Brodt S (2003) Beyond the local/global divide: knowledge for tree management in Madhya Pradesh. In: Sivaramakrishnan K, Agrawal A (eds) Regional modernities. The cultural politics of development in India. Stanford University Press, CA, pp 338–358

Brunsson N, Olsen J (1998) Organizing organizations. Fagbokforlaget, Bergen

Chambers R, Pacey A, Thrupp LA (eds) (1989) Farmer first: farmer innovation and agricultural research. Intermediate Technology Publications, London

Cohen M (1983) Learning by doing: World Bank lending for urban development, 1972–82. World Bank, Washington DC

Critchley WRS (2000) Inquiry, initiative and inventiveness: farmer innovators in East Africa. Phys Chem Earth Part B 25(3):285–288

Dattée B, Birdseye Weil H (2007) Dynamics of social factors in technological substitutions. Technol Forecast Soc Chang 74:579–607

de Certeau M (1990) L'invention du quotidien. I Arts de faire. Gallimard, Paris (Italian translation: L'invenzione del quotidiano. Edizioni Lavoro, Roma, 2001)

Dosi G (1982) Technological paradigms and technological trajectories: a suggested interpretation of the determinants and directions of technical change. Res Policy 11:147–162

Elzen B, Wieczorek A (2005) Introduction: Transitions towards sustainability through system innovation. Technol Forecast Soc Chang 72:651–661

Emmerij L (2002) Aid as a flight forward. Dev Chang 33(2):247–260

Escobar A (1995) Encountering development: the making and unmaking of the third world. Princeton University Press, Princeton

Escobar A (1996) Constructing nature: elements for a post-structural political ecology. In: Peet R, Watts M (eds) Liberation ecologies. Environment, development, social movements. Routledge, London, pp 46–68

Escobar A (1998) Whose knowledge, whose nature? Biodiversity, conservation, and the political ecology of social movements. Journal of Political Ecology 5:53–82

Escobar A (2001) Culture sits in places: Reflections on globalism and subaltern strategies of localization. Political Geography 20:139–174

Feinberg Y, Stewart C (2008) Testing multiple forecasters. Econometrica 76(3):561–582

Ferber J (1999) Multi-agent systems: an introduction to distributed artificial intelligence. Addison-Wesley-Longman, Harlow

Forester J (1999) The deliberative practitioner. MIT Press, Cambridge, MA

Galimberti U (1999) Psiche e Teche (Psyche and Techne). Feltrinelli, Milan

GEC–Gujarat Ecology Commission (2002) Reconciling water use and environment. Water resources management in Gujarat: Resources, problems, issues. strategies and framework for action. Sectoral report on hydrologic regimes subcomponent, GEC, Vadodara

Geschiere P, Meyer B (1998) Globalization and identity. Dialectics of flow and closure. Development and Change 29:601–615

Gifford A (2005) The role of culture and meaning in rational choice. J Bioecon 7:129–155

Gilbert AG (1994) Third World cities: poverty, employment, gender roles and the environment during a time of restructuring. Urban Studies 31(4/5):605–633

Grassini L (2002) Spazi di interazione tra locale e globale: Pianificazione e pratiche di gestione delle risorse idriche in India (Interactions between local and global: Water planning and management practices in India). PhD thesis. University La Sapienza, Rome

Grassini L (2007) Local-Global interactions: Silent practices of change in a slum upgrading project in India", paper presented at the AESOP Conference, Naples, July 10-14, 2007

Grover B (1998) Twenty-five years of international cooperation in water-related development assistance, 1972–1997. Water Policy 1:29–43

Guha R (2000) Environmentalism. A global history, Longman, New York

Gupta A (1998) Respecting, recognising and rewarding local creativity: knowledge networks for biodiversity conservation and natural resource management, Indian Institute of Management, Ahmedabad (mimeo)

Henderson R, Clark KB (1990) Architectural innovation: the reconfiguration of existing. Adm Sci Q 35(1):9–30

Hirschman AO (1967) Development projects observed. The Brookings Institution, Washington DC [Italian translation: I progetti di sviluppo, FrancoAngeli, Milano, 1975]

Hobart M (1993) Introduction: The growth of ignorance? In: Hobart M (ed) An anthropological critique of development. the growth of ignorance. Routledge, London, pp 1–30

Kemp R, Rotmans J (2004) Managing the transition to sustainable mobility. In: Elzen B, Geels FW, Green K (eds) System innovation and the transition to sustainability. Edward Elgar, Cheltenham, pp 137–167

Kessides C (1997) World Bank experience with the provision of infrastructure services for the urban poor: preliminary identification and review of best practices. World Bank, Washington DC

Ketelar T, Todd P (2001) Framing our thoughts: ecological rationality as evolutionary psychology's answer to the frame problem. In: Holcomb HR (ed) Conceptual Challenges in Evolutionary Psychology: Innovative Research Strategies. Kluwer, Norwell, MA, pp 179–211

Kloppenburg J (1991) Social theory and the de/reconstruction of agricultural science: Local knowledge for an alternative agriculture. Rural Sociology 56:519–548

Lanzara GF (1993) Capacità negativa. Competenza progettuale e modelli di intervento nelle organizzazioni (Negative capacity. Planning competencies and intervention models in organizzazioni). Il Mulino, Bologna

McCully P (1996) Silenced rivers: the ecology and politics of large dams. Zed Books, London

Mignolo W (2000) Local histories/global designs: Coloniality, subaltern knowledges, and border thinking. Princeton University Press, Princeton

Millar J, Curtis A (1999) Challenging the boundaries of local and scientific knowledge in Australia: Opportunities for social learning in managing temperate upland pastures. Agric Hum Values 16:389–399

Mishra A (1994) Rajasthan Ki Rajat Bundein. Gandhi Peace Foundation, New Delhi (English translation: The radiant raindrops of Rajastan, Research Foundation for Science, Technology and Ecology, New Delhi, 2001)

Moore GA (2002) Crossing the chasm: Marketing and selling high-tech products to mainstream customers. Harper&Collins, New York

Norton JA, Bass FM (1992) Evolution of technological generations: the law of capture, Sloan Management Review, Winter: 66-77

O'Hare GN, Jennings NR (1996) Foundations of Distributed Artificial Intelligence. Wiley, New York

Okamoto S, Sycara K, Scerri P (2009) Personal assistants for human organizations. In: Dignum V (ed) Organizations in multi-agent systems. IGI-Global Handbook of Research, Hershey, PA, pp 514–540

Parikh H (1995) Slum networking. A community-based sanitation and environmental improvement programme: experiences from Indore, Baroda and Ahmedabad. Report, Human Settlement Management Institute New Delhi, and Institute for Housing and Urban Development Studies, The Netherlands

Parikh KS (1998) Poverty and environment. Turning the poor into agents of environmental regeneration. UNDP Working Paper 1

Peet R, Watts M (1996) Liberation ecologies. Environment, development, social movements. Routledge, London

Pendharkar PC (2007) The theory and experiments of designing cooperative intelligent systems. Decis Support Syst 43:1014–1030

Postel S (1998) Pillars of sand: Can irrigation miracle last?. Norton, New York

Prasad E (2001) Emulating civil society. In: Agarwal S, Narain S, Khurana I (eds) Making water everybody's business. Practices and policy of water harvesting. Centre for Science and Environment, New Delhi, pp 364–382

Ritzer G (2003) Rethinking globalisation: glocalization/globalization and something/nothing. Sociological Theory 21(3):193–209

Rogers EM (2003) Diffusion of innovations. The Free Press, New York

Roy A (2002) Guerra è pace. Ugo Guanda, Parma

Schön DA (1973) Beyond the Stable State. Penguin, Harmondsworth

Severino E (1998) Il Destino della Tecnica (Technique and its Destiny). Rizzoli, Milan

Shah T (2000) Mobilising social energy against environmental challenge: Understanding the groundwater recharge movement in western India. Natural Resources Forum 24:197–209

Shakun M (1999) Consciousness, spirituality and right decision/negotiation in purposeful complex adaptive systems. Group Decision and Negotiation 8:1–15

Shiva V (2001) Water wars. Privatisation pollution and profit. Pluto Press, London

Shiva V (2002) Privatisation, pollution and profit. South End Press, Cambridge, MA

Shiva V, Bandyopadhyay J (1990) Asia's forests, Asia's cultures. In: Head S, Heinzman R (eds) Lessons of the rainforest. Sierra Club Books, San Francisco

Sivaramakrishnan K, Agrawal A (2003) Regional modernities: The cultural politics of development in India. Stanford University Press, Stanford and Oxford University Press, Delhi

Sridhar A (2001) Study, documentation and assessment of the feasibility of revival of traditional drinking water harvesting systems and practices prevalent in the state of Gujarat. Report, Pravah, Ahmedabad

Swyngedouw E (1992) Territorial organization and the space/technology nexus. Transactions Institute of British Geographers NS 17:417–433

Swyngedouw E (1997) Neither global nor local: 'glocalisation' and the politics of scale. In: Cox K (ed) Spaces of Globalization. Guilford Press, New York, pp 137–166

Tripathi D (1998) Alliance for Change, A Slum Upgrading Experiment in Ahmedabad. Tata McGraw-Hill, New Delhi

van de Kerkhof M, Wieczorek A (2005) Learning and stakeholder participation in transition processes towards sustainability: Methodological considerations. Technological Forecasting & Social Change 72:733–747

Vattimo G (1997) Tecnica ed esistenza (Technique and existence). Paravia, Turin

Weick KE (1995) Sense making in organizations. Sage, London

Weick KE (2001) Making sense of the organisations. Blackwell, Oxford

World Bank (1998) Assessing aid: what works what doesn't and why. Oxford University Press and World Bank, New York

World Bank (2001) World Development Report: attacking poverty. Oxford University Press and World Bank, New York

Worster D (1985) Nature's economy. A history of ecological ideas. Cambridge University Press, Cambridge

WSSCC (1999) Water supply and sanitation collaborative council vision 21. A shared vision for water supply, sanitation and hygiene and a framework for future action. Report, WSSCC

Chapter 5
Surface Runoff and Sediment Yields from Tropical Volcanic Upland Watersheds as Influenced by Climatic, Geological and Land-Use Factors

Anton Rijsdijk

Abstract This paper presents discharge, suspended sediment and bedload yields from a young volcanic catchment (the Konto river catchment) in East Java, Indonesia. river flow and catchment sediment yields were determined in three sub-catchments at two locations in each. In general, the upper stations represented more-or-less natural conditions, while two of the lower stations included agricultural land and settlements. Two sub-catchments, which had a predominantly forested upper part, and a lower catchment with settlements and a high percentage of irrigated or rain-fed agricultural land, showed clear differences in quickflow percentages and suspended sediment yield. In these sub-catchments, quickflow (expressed as a percentage of rainfall) at the lower stations was twice as much as that observed in the forested upper stations, while the annual suspended sediment load was between three times higher (3.4 vs. 11.2 Mg ha^{-1}) and 26 times higher (0.35 vs. 9.3 Mg ha^{-1}). By contrast, in another sub-catchment with little agriculture and no settlements, there was virtually no difference between the upper and lower station (both around 1.0 Mg ha^{-1}). The contribution of bedload to the total sediment load ranged from 0.5% in an upper watershed with consolidated material to nearly 12% in an upper watershed situated in an unstable lahar valley. In both developed lower catchments, the bedload component was about 6.5% of the total load. From erosion-pin measurements and analysis of the grain-size contribution of both bed material and the riverbank material, the contribution of bank erosion was estimated to range from 1 to about 36% of the total sediment load. River sediment yields in Indonesia appear to be underestimated due to incorrect sampling and calculation methods. Comparison between the conventional method and

---

Anton Rijsdijk—Formerly With DHV Consulting Engineers, Amersfoort, The Netherlands

---

A. Rijsdijk (✉)
J. van der Borchstraat 24bis, 3515 XE, Utrecht, The Netherlands
e-mail: rijsdijk.1@inter.nl.net

D. Higgitt (ed.), *Perspectives on Environmental Management and Technology in Asian River Basins*, SpringerBriefs in Geography, DOI: 10.1007/978-94-007-2330-6_5, © The Author(s) 2012

theoretically more accurate approach yielded an underestimate of up to 47% of the first method. However, despite the different methods and uncertainties, the results obtained in this study are comparable with those of other studies on sediment yield in the humid tropics.

Keywords Humid tropics · Sediment yield · Water balances · Bank erosion · Bedload · Sediment measurements

5.1 Introduction

The high sediment content of the rivers in Indonesia and neighbouring countries has been considered to be a serious problem since the first publication on this subject nearly a century ago (Rutten 1917). Since then, many more recent studies have been conducted on hydrology and sediment yield in tropical watersheds, including those by Van Dijk and Ehrencron (1949), Bruijnzeel (1983) and Van der Linden (1978) on Java, by Amphlett (1988) in the Philippines, by Balamurugan (1991) in Malaysia, and by Turkelboom (1999) in Thailand. Nevertheless, many questions about the scale of the problem and the nature of sediment transport still remain unanswered.

One of the assumptions, which was taken for granted during many years of erosion research in Indonesia was: "Deforestation causes surface erosion and loss of agricultural productivity". While this is not entirely untrue, and some studies show a clear correlation between deforestation and increasing sediment yield [e.g. in the Cilutung tributary, as cited by Ambar and Wiersum (1980], developments in research show that the issue is quite complicated and some nuances might be opportune. As Diemont et al. (1991) point out, low inputs, not surface erosion, are the causes of low productivity. He argues that if rain-fed agriculture was the single cause of the high sediment yield in rivers, then all the efforts to reduce erosion in the last 50 years, should have paid off. Besides the reduction in erosion due to terracing [estimated by Diemont et al. (1991), to have been implemented in 75% of all cultivated uplands in Java], the erosion should also have been reduced by the ongoing conversion of rain-fed agricultural land to irrigated rice fields in Java, even if the time lag in sediment delivery is taken into account. However, despite all of this, the river sediment loads remain high and sedimentation continues to be problematic. Hence, a comprehensive study of all possible sediment sources in a watershed, linked to the basin sediment output, was therefore essential to understand the influences of deforestation on the river sediment yield at different scales.

The Konto river watershed management project near the East Java town of Malang in Indonesia, which was implemented in response to the rapid siltation of the artificial Lake Selorejo (Brabben 1979), provided an opportunity to investigate the relationship between land use and sediment yield in detail.

This project, which started in 1979, was a joint multidisciplinary study conducted by the governments of the Netherlands and Indonesia. It aimed to develop the planning procedures needed to establish a management model for the forested land which occurred in densely populated watersheds in Java. Within the framework of this project (Anonymous 1989), attempts were made to quantify the sediment yield of the contributing rivers, to analyse the hydraulic behaviour of the watershed, and to evaluate the (often strongly interrelated) sediment sources (Rijsdijk and Bruijnzeel 1990, 1991). The main erosion research activities were carried out from 1987 until the end of the project in 1990. The present paper summarises the main results of a recent re-analysis of the original data on river flow and sediment yield as collected by the Konto river project.

The main objective of the study presented here was to investigate the effect of the different sub-catchments' hydrology (water balance and quickflow) on sediment yield (suspended load and bedload). The study also aimed to present accurate data on sediment yield for planning purposes, provide increased insight into the hydrological behaviour and sediment transport of upland volcanic catchments, and contribute to the discussion on the 'true' sediment yield of rivers. Other studies on erosion and sediment dynamics of the Konto river basin have considered the runoff and sediment production from rural roads, trails and settlements (Rijsdijk et al. 2007a), the sediment yield from gullies, riparian mass wasting and bank erosion (Rijsdijk et al. 2007b), and the evaluation of sediment sources and delivery (Rijsdijk 2005).

An important issue is the discussion on the 'true' sediment yield of the Indonesian rivers. In Indonesia, the suspended sediment flux is usually calculated using the classical sediment rating curve–flow duration method (Miller 1951). In this, samples are lumped together and converted to stream sediment load using a stage-suspended-sediment load relation curve (Q–S) for each station. This is done for all seasons and years together, as recommended by Jahani (1992). In addition, sampling in Indonesia is often only conducted during office hours, although most floods occur late in the afternoon or evening.

However, several investigators have reported serious underestimation of sediment loads when the sediment rating curve–flow duration method is applied in small or medium-sized basins (Walling 1977a, b; Ferguson 1986; Walling et al. 1992), even when data were corrected for the bias caused by log transformation (Ferguson 1986; Walling and Webb 1988; Thomas 1988). Hence it is relevant to compare the conventional values with theoretically more correct methods, such as the event-based method (Guy and Norman 1970; Bruijnzeel 1983).

This problem of under estimating the sediment load is certainly not only an Indonesian problem, For example, White (1988) found that on the average, the predicted sedimentation rate was only 39% of the observed rate measured at ten reservoirs in India, Indonesia, Kenya and the Philippines.

5.2 Study Area

The Konto river is a part of the large Brantas river system (Fig. 5.1) and drains a young volcanic watershed of 233 km^2. The watershed is located 25 km NW of the East Java town of Malang and had a total population of about 100,000 in 1990.

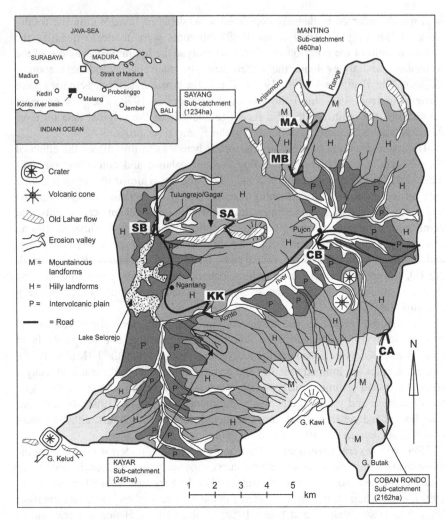

Fig. 5.1 Location and general geomorphology of the Upper Konto catchment, East Java (after Nuffic-Unibraw 1984), with locations of study sites added (based on Rijsdijk 2005). CA = Coban Rondo upper station, CB = Coban Rondo lower station; MA = Manting upper station, MB = Manting lower station; SA = Sayang upper station, SB = Sayang lower station; KK = Konto river at the village of Kambal

It comprises intervolcanic plains and plateau's (25% of the total area), alluvial and lahar valleys (5%), hilly areas (50%) and volcanic mountain complexes (20%). Within the Konto watershed, three sub-catchments were investigated in detail. These catchments comprise about 17% of the total area and include the Coban Rondo river catchment in the south-eastern part, the Manting sub-catchment in the north and the Sayang catchment in the north-west corner of the Konto river watershed. The Coban Rondo catchment (2,162 ha) is situated on the northern slopes of the Kawi–Butak volcanic massif. The upper part, from 2,600 m down to about 1,700 m is quite steep and relatively inaccessible while the lower hilly part slopes gently to 1,100 m at the confluence with the Konto river. The much smaller Manting sub-catchment (460 ha) is bordered to the north by the inner slopes of the Anjasmoro volcano range and slopes down from 1,900 to 1,250 m. The upper Sayang catchment (1,233 ha) is partly located in a lahar valley bordered to the east by the Anjasmoro mountains at 1,600 m and slopes down to the inflow of the Sayang river into Lake Selarejo at 700 m ASL. As in the Manting and Coban Rondo basins, slopes in the upper part of the Sayang basin are steep but gradually give way to hilly landforms and intervolcanic plains at lower altitudes.

The climate of the study area is classified according to Oldeman (1975) as agroclimate 'C2'. This climate, which is typical for higher elevations in a tropical monsoon climate, has relatively pronounced wet seasons (December–April) and dry seasons (June–October). The mean annual rainfall is about 2,400 mm (Brantas 1989), based on long-term records (1917–1988) from the rain gauges in and around the project area. However, great spatial variation in rainfall exists. Intense rainstorms can also occur: the highest daily rainfall recorded during the study period was 88 mm in the Coban Rondo sub-catchment, 84 mm in the Manting sub-catchment and 102 mm in the Sayang sub-catchment (Rijsdijk and Bruijnzeel 1990).

All soils in the upper Konto area have developed from windblown, fine-textured volcanic ash and are classified (FAO 1990) as Andosols (mountainous parts), Cambisols (lower slopes and foothills) and Luvisols (lower plains with wetland rice cultivation). More details on the soil properties in the Konto river basin are presented in Nuffic-Unibraw (1984).

Table 5.1 presents the land use in the Konto river area. About two-thirds of the area—in general the part above 1,400 m—is covered with (degraded) natural forest or tree plantations. At lower altitudes, the degraded natural forests are gradually turning into shrubland. This economically unproductive shrubland, in turn, is often replaced by agroforestry (*tumpangsari*). The lowest parts of the project area are used for wetland rice cultivation or for rain-fed agriculture with non-perennial crops or coffee; these areas are often also occupied by human settlements, roads and trails. More details about the geography of the project area and the project can be found in RIN (1985) and Rijsdijk and Bruijnzeel (1990).

5.3 Monitoring Programme

5.3.1 Water Balance and Sediment Discharge Measurements

From 1988 to 1990, rainfall data were collected daily and at 15-min intervals using various types of rain gauges. These ranged from manual daily rain gauges and mechanical tipping bucket rain gauges (SIAP, Bologna, Italy) in the lower part of

Table 5.1 Areas of major types of land use in 1989 in the entire Konto river watershed (East Java, Indonesia), and in the upper (A) and lower (B) parts of three sub-catchments within the Konto watershed. see Fig. 5.1

Unit	CA		CB		MA		MB		SA		SB		Konto	
	ha	%	ha	%	ha	%	ha	%	ha	%	ha	%	ha	%
(Disturbed) natural forest	1,034	88	1,049	49	318	82	318	69	205	60	437	35	5,700	24
Plantation forest	3	0	142	7	4	1	25	6	57	17	149	12	1,200	5
Tumpangsari			90	4	1	0	31	7	10	3	11	1	800	3
Scrub	133	11	172	8	65	17	85	19	63	18	292	24	7,000	30
Bamboo			62	3							6	0	500	2
Coffee and mixed garden			9	0					1	0	53	4	800	3
Rainfed agricultural land			333	15					4	1	95	8	4,000	17
Irrigated rice fields (sawah)			166	8							164	13	2,200	9
Settlement			138	6							27	2	800	3
Lake													300	1
Roads													300	
Total area (ha)	1,170	100	2,162	100	388	100	460	100	341	100	1,233	100	23,300	100
Altitude of gauging station (m ASL)	1,700		1,100		1,350		1,250		1,050		700		750 (Kambal)	

C Coban Rondo, *M* Manting, *S* Sayang

the basins, to electronic long-term recording devices (Obsermet, Ridderkerk, The Netherlands) at the top of the mountains. Due to frequent malfunctions of the long-term rain gauges, the data from higher altitudes are under-represented. All measured rainfall data were converted to catchment area rainfall using the Thiessen polygon method. For comparison, data from the existing meteorological stations in Pujon and Ngantang were collected (Brantas 1989).

The catchment evapotranspiration (ET) was calculated as a weighted mean of the evapotranspiration values of the dominant types of landuse. The (Penman) evaporation (Eo) values were computed from meteorological data at the project's meteorological station at lower Manting (Rijsdijk and Bruijnzeel 1990) and corrected for altitude by equations developed for central Java (Bruijnzeel 1988). Next, the evapotranspiration rates of the dominant types of landuse were calculated according (ET = crop factor *Eo). Details on the methodology and crop factors are in Rijsdijk and Bruijnzeel (1990).

From 1988 to 1990, river discharges and river sediment load (suspended load and bedload) were measured at two locations within each of the three sub-catchments: Coban Rondo (C), Manting (M) and Sayang (S). The upper ('A') gauging stations (CA, MA and SA) were located just below the (relatively) undisturbed and steep parts of the sub-catchments; the lower ('B') stations (CB, MB and SB) were located at the outflow of each sub-catchment (Fig. 5.1). The suspended sediment yield of the Konto river, at the Kambal station (KK, Fig. 5.1) was assessed in 1990.[1] However, these results are only indicative (Rijsdijk and Bruijnzeel 1991).

Water levels were recorded using hydraulic measuring devices equipped with automatic water-level recorders (Ott, Kempten, Germany). Due to mechanical problems and damage by large floods, the 1990 outflow data from stations CB and SB are probably somewhat underestimated, while the MA station was completely destroyed by an extreme flood in January 1990. Stage discharge (Q–H) relations were established using the area velocity method and data from current meters (Ott, Kempten, Germany) and converted to daily and annual values. The wet season started usually in November, with low inter annual variations in the October flow,[2] hence the period from November to December was taken as a hydrological year.

5.3.2 Suspended Load Sampling Procedures

At each gauging station, hundreds of grab samples of water and suspended sediment were taken during the wet season, both in storms and during periods of baseflow (Table 5.2). During the rising stage of a storm, flow samples were taken at intervals of a 1- or 2-cm increase in water level or with time intervals of up to

---

[1] Using the sediment rating curve–flow duration method (SRC–FDM).

[2] For example, the highest inter annual variation (at station MA between 1988 and 1989) was less than 8% of the monthly maximum flow in 1989.

Table 5.2 Number of samples and concentrations of suspended sediment in the rivers measured at the gauging stations of the studied sub-catchments

	No. of samples		Sediment concentration (g l⁻¹)		
	Base flow	Storm flow	Min	Max	Median
CA	16	94	0.003	0.985	0.068
CB	25	88	0.067	13.792	2.87
MA	33	67	0.002	0.552	0.074
MB	16	60	0.004	0.929	0.031
SA	28	122	0.007	15.198	0.396
SB	35	133	0.103	22.086	1.554

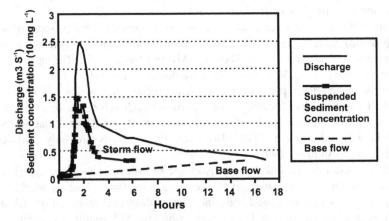

Fig. 5.2 Example of a storm event analysis. Broken line indicates stormflow separation line according to the method of Hewlett and Hibbert (1967)

15 min. Most sampling sessions lasted at least until the summit of the flood wave was reached and often included a major part of the falling limb as well (Fig. 5.2). All samples were taken downstream of the bedload trap at the crest of the weir. This sampling method can be used as a reasonable substitute for depth integrated sampling [Østrom (1975), quoted in Bogen (1992)]. The water and sediment samples were filtered with pre-weighted Melita filters,[3] dried for 24 h at 105°C and weighed to the nearest 0.001 g.

---

[3] This methodology could result in slight under-estimation of the sediment yield in comparison to filtration through 0.45 μm Millipore filters. Purwanto (1999), in west Java, compared the two methods and his results suggested ca. 4.5% lower values for the paper filters.

Table 5.3 Statistical parameters of regression equations linking rainfall (P) to quickflow (Qq) per event at the studied sub-catchments

Site	No. of observations	Range (mm)	R^2	SEE	F-value	P level
CA	19	7–74	0.27*	19	6.3	<0.1
CB	46	0.3–61	0.27*	13.7	16.4	<0.001
MA	24	4–51	0.63	6.8	38.2	<0.001
MB	19	4–51	0.75	5.7	50.5	<0.001
SA	68	0.14–79	0.55	10.7	81.5	<0.001
SB	55	0.31–70	0.64	9.2	94.8	<0.001

5.3.3 Calculation Procedures

Baseflow and quickflow were separated by drawing a line from the point on the hydrograph at which the water level started to rise at an angle of 3.76E-05 m³/s. The area above the line was considered to be quickflow and that below this line baseflow (Hewlett and Hibbert 1967). Quickflow volumes per storm were regressed against corresponding rainfall for each station. Daily average rainfall records for each station were inserted into the regression equation to determine the annual amount of quickflow.

These relations were in general significant at $p < 0.001$, with the exception of station CA ($p < 0.1$). The low coefficient of determination ($R^2 = 0.27$) of the Coban Rondo stations, (CA and CB) were probably caused by the frequent malfunctions of the electronic rain gauges at higher altitudes (Table 5.3).

The yield of suspended sediment was calculated using the event-based method (Guy and Norman 1970; Bruijnzeel 1983). Using this method, the sediment yield per storm was calculated individually and related to the corresponding amount of quickflow. When a number of stormflow events were sampled, a power curve linking log quickflow volume and log sediment yield was established. These relations were significant with a $p < 0.001$, except CA ($p < 0.01$ and CB ($p < 0.1$) (see Table 5.4 and Fig. 5.2). Using this relationship, the total sediment yield was estimated by calculating the quickflow volume of each event from the runoff hydrograph and adding the sediment volume of the baseflow to this (details in Kaatee 1989).

5.3.4 Bedload Measurements

Bedload was measured using large concrete slot traps.[4] Although this method avoided the problems in measurement created by portable samplers (Emmett 1981;

---

[4] In rivers with low bedload yield the traps were placed perpendicular to the river flow and covered 100% of the river bed, while in high yielding rivers multiple traps were placed parallel and covered 40–75% of the river bed. The total volume ranged from 0.75 M³ (MA) to 5.25 M³ (SB).

Table 5.4 Statistical parameters of regression equations linking log transformed quickflow (log QF) to log transformed sediment yield (log SY) per event at the studied sub-catchments

Station	Number of storms	log QF vs. log SY			
		R2	SEE	F-value	P level
CA	9	0.79	0.28	25.6	<0.01
CB	18	0.84	0.19	86.1	<0.001
MA	6	0.8	0.36	20.9	<0.1
MB	7	0.92	0.14	60.2	<0.001
SA	14	0.82	0.46	54.4	<0.001
SB	13	0.89	0.25	95.6	<0.001

Ward 1984b), slot traps introduced errors of their own. All traps were measured and cleaned daily, but some bedload traps (at the SB station and to a lesser extent at the CB station) overflowed occasionally during large storms, causing the bedload volume to be underestimated. On the other hand, when observers failed to clean the traps thoroughly, the residue in the traps was added to the results of the following day, causing overestimation of the amount of bedload. The volume was converted to weight using a conversion factor of 1.1 (Rijsdijk and Bruijnzeel 1990).

5.3.5 Bank Retreat

In the wet season of 1988/1989, bank retreat rates were measured by means of erosion pins driven into the riverbank both at actively eroding bends and at straight sections of the riverbanks of the Sayang (four sites) and at three sites at the Coban Rondo rivers (for full details of methods see Rijsdijk and Bruijnzeel 1990). The results of the measurements are only indicative (error margins at least 50%), as bank erosion is very irregular, while pins may alter the streamflow along the bank or could influence the stability of the riverbank (Thorne 1981; Kaatee 1989).

In addition to this method, grain-size analyses of both bed material and riverbank material were used to correlate the volume lost through bank erosion with the amount of sediment caught in the bedload traps. Representative soil samples (eight in Coban Rondo, six in Manting and eight in Sayang) were taken from the riverbank, and samples were also taken of the material in the bedload traps and of the accumulated bed sediment in the river channels. Grain size was analysed and the ratio of gravel (fraction > 2 mm) in the bank and bed material was derived. From this, assuming that all gravel originated from the banks, and not from sheet erosion,[5] the amount of bank erosion was estimated. Again, results should only be considered indicative, with estimated error margins at least 50%.

---

[5] The values of CB and SB have been corrected for the contribution of roads (Rijsdijk and Bruijnzeel 1990).

Table 5.5 Water balances of the upper (A) and lower (B) parts of three sub-catchments within the Konto watershed

	Year[a]	CA	CB	MA	MB	SA	SB	Pujon	Ngantang
Annual rainfall (mm)	1988	2,201	1,968	2,347	2,292	2,135	2,195	1,947	2,001
	1989	2,545	2,267	3,091	3,227	2,290	2,351	1,989	2,487
	1990	2,110	2,416	4,478	4,349	2,961	3,264		
	Average[b]	2,373	2,118	2,719	2,760	2,213	2,273	1,968	2,244
	Average[c]	2,285	2,217	3,305	3,289	2,462	2,603	2113[d]	2547[d]
Annual evapotranspiration (mm)	Average	700	740	825	820	935	975		
Annual runoff (mm)	1988	783	612	2,259	2,009	1,079	1,170		
	1989	1,023	926	2,300	2,209	1,401	1,295		
	1990	970	435[e]	[f]	2400[g]	1,550	975		
	Average[b]	903	769	2,280	2,109	1,240	1,233		
Average annual water balance (mm)		770	609	−386	−170	38	66		

C Coban Rondo, *M* Manting, *S* Sayang
[a] 1988 = November 1987–October 1988; 1989 = November 1988–October 1989; 1990 = November 1989–April 1990
[b] 2-year average (1988–1989)
[c] 3-year average (1988–1990)
[d] 1950–1985 (Pujon) and 1954–1985 (Ngantang)
[e] Values probably underestimated
[f] Station destroyed by large flood
[g] Data unreliable or estimated

5.4 Results

5.4.1 Hydrology

Table 5.5 presents the 3-year annual[6] areal rainfall, evaporation, runoff and water balance for the six sub-catchments during the period November 1987–April 1990. A comparison of annual rainfall data of each sub-catchment with the long-term averages of data recorded at the meteorological stations at Pujon (2,113 mm) and Ngantang (2,547 mm) (Brantas 1989), reveals that the hydrological year 1987/ 1988 was relatively dry (2,152 mm rainfall), 1988–1989 was more or less normal (2,615 mm) and 1989–1990 was relatively wet (3,343 mm). Rainfall averages for the three catchments yielded clear differences Manting 3,289 mm, Coban Rondo 2,217 mm and 2,603 for Sayang (Table 5.5). In January 1990, extreme events occurred in Manting: in that month 1,594 mm rain fell, resulting in a minimum outflow of 770 mm.[7] Unfortunately, due to the malfunction of equipment, daily

---

[6] Hydrological year (November–October) for 1987–1988 and 1988–1989 (November–April for 1989–1990).
[7] Probably underestimated due to damage to the equipment.

Table 5.6 Runoff coefficients (%), quickflow (as a proportion of rainfall and total flow) and contributing areas of the upper (A) and lower (B) parts of three sub-catchments within the Konto watershed

Year[a]	CA	CB	MA	MB	SA	SB
RC % (1988)	36	31	96	88	51	53
RC % (1989)	40	41	74	68	61	55
RC % (1990)	46	18	[b]	55	52	30
RC Average[c]	38	36	85	78	56	54
Qq/P %	1.6	3.3	5.9	5.7	4.9	9.8
Qq/Qt%	4.2	9.0	7.0	7.5	8.8	18.0
MCA	19	71	22	26	17	121
HOF/SOF contributing area[d]	2	308	1	2	5	195

C Coban Rondo, *M* Manting, *S* Sayang
MCA = Minimum contributing area
HOF = Horton Overland Flow
SOF = Saturated Overland Flow
[a] 1988 = November 1987–October 1988; 1989 = November 1988–October 1989; 1990 = November 1989–April 1990
[b] Station destroyed by large flood
[c] 2-year average (1988–1989)
[d] Estimate from land use, see text

rainfall records are not available for this period. No data are available for MA as this station was completely washed away during the first flood.

As explained in Sect. 3.1, the annual water yields recorded for 1989–1990 for the MB, MA and SA station are less reliable, hence all water balance calculations are derived from the 1987–1988 to 1988–1989 data. The overview of water balances (Table 5.5) shows remarkable differences between the three sub-catchments. The upper and lower stations in Coban Rondo (CA and CB) indicate a net loss of 770 and 609 mm, respectively, while the Manting stations (MA and MB) show net gains of 386 and 170 mm, respectively; Sayang (SA and SB), appears to be more or less in equilibrium (loss of 36 and 66 mm, respectively). The results on quickflow (Qq/Qt%) compare very well with those of other investigators in similar watersheds with deep and permeable soils in SE Asia (Sect. 5.2 and Tables 5.6 and 5.7)

5.4.2 Sediment Yields

The suspended sediment yields of all catchments, calculated using the event-based method (Sect. 3.3), are shown in Table 5.8. The sediment values from 1990 were not used to calculate the averages as they are considered less reliable. The high suspended sediment yield obtained for Manting for 1989–1990 (5.48 Mg ha^{-1}) is not an error, but is the result of extremely heavy rainfall in January 1990. For example, on January 25 and 26, the yield was as high as 1 Mg ha^{-1} each day.

Table 5.7   Runoff coefficient (Qq/Qt%) for selected tropical rainforest catchments in SE Asia

Catchment	Country	Size (km^2)	Qq/Qt %	Separation technique			Source
W8S5	Malaysia	1–7	51	N-days after peak	Forest	Shallow soil	Biden and Greer (1977)
W8S5	Malaysia	1–7	41	Inclined line	Forest	Shallow soil	Biden and Greer (1977)
Skudai River,	Malaysia	0.2	22	Inclined line	Plantation forest	Sandy clay	Chong et al (2006)
Sg Air Terjun	Malaysia (Pinang Island)	4.8	15	Unknown	Forest	?	Ismail (1997)
Sg Relau	Malaysia (Pinang Island)	11.2	22	Unknown	Developed	?	Ismail (1997)
Kali Mondo	Indonesia	0–19	5–7	Chemical tracer	Mainly forest and shrub	Deep volcanic	Bruijnzeel (1982)
Cikumutuk	Indonesia	1	14–16	Inclined line	70% forest and shrub	Deep volcanic	Purwanto (1999)
Kali Konto	Indonesia (East Java)	3–22	4–18	Inclined line	64–100% forest and shrub	Deep volcanic	This study

Separation techniques:
Inclined line (Hewlett and Hibbert 1967) N-days after peak (Linsley et al 1975) Chemical tracer (Bruijnzeel 1982)

In addition, Table 5.8 presents the annual bedload volumes which, like the suspended sediment values, are highly variable. The bedload values at the SB station for 1987–1988 are probably an underestimate as the bedload traps over-flowed frequently that year.

Estimates obtained through the riverbank erosion-pin study revealed that, in 1988–1989, the lower Sayang river widened by approximately 8 cm and the Coban Rondo lower river widened by approximately 4 cm. A comparison of the grain-size distribution of the riverbank material, the material in the traps and loose sediment on the riverbed, revealed that 30% of the bank material from Coban Rondo and Sayang ends up as bedload, while 50% of that from Manting ends up as bedload. Using the grain-size distribution method, the following channel erosion retreat rates were estimated: Coban Rondo upper river 0.2 cm (0.02 Mg ha^{-1} or 5% of the total load); Coban Rondo lower river 6 cm (1.8 Mg ha^{-1} or 18% of the total load); Manting upper river 0.01 cm (0.01 Mg ha^{-1} or 1% of the total load); Manting lower river 0.3 cm (0.08 Mg ha^{-1} or 7% of the total load); Sayang upper river 11 cm (1.4 Mg ha^{-1} or 36% of the total load); Sayang lower river 7 cm (2.0 Mg ha^{-1} or 17% of the total load). Again these results are only indicative.

Table 5.8 Annual sediment load (Mg/ha) of the upper (A) and lower (B) parts of three sub-catchments within the Konto watershed

	Year[a]	CA	CB	MA	MB	SA	SB
Suspended sediment	1988	0.23	5.23	0.77	0.81	3.41	11.79
	1989	0.48	13.32	1.13	1.25	3.40	10.50
	1990	0.41[b]	6.97[c]	–[d]	5.48[e]	7.15[b]	6.21[c]
	Average (1988–1989)	0.35	9.27	0.95	1.03	3.40	11.15
Bedload	1988	0.009	0.54	0.005	0.04	0.49	0.35[c]
	1989	0.004	0.78[f]	0.004	0.04	0.43	0.74
	Average (1988–1989)	0.01	0.66	0.005	0.04	0.46	0.74[g]
Bedload as % of total load	Average (1988–1989)	1.8	6.7	0.5	3.6	11.9	6.3[g]
Total sediment load	1988	0.24	5.77	0.77	0.85	3.90	12.14[c]
	1989	0.48	14.10	1.13	1.29	3.83	11.24
	Average (1988–1989)	0.4	9.9	1.0	1.1	3.9	11.7

C Coban Rondo, *M* Manting, *S* Sayang
[a] 1988 = November 1987–October 1988; 1989 = November 1988–October 1989; 1990 = November 1989–April 1990
[b] Dry season not included
[c] Values probably underestimated
[d] Station destroyed by large flood
[e] Data unreliable or estimated values
[f] Values probably overestimated
[g] Calculated from 1989 only

5.5 Discussion

5.5.1 Water Balance

Orographic and exposure effects could explain the higher rainfall totals obtained for the Manting sub-catchment (the average for Manting was 3,289 mm vs. 2,217 mm for Coban Rondo and 2,603 mm for Sayang). However, the frequent breakdowns of the electronic raingauges on top of the mountains excluded any analysis of the influence of altitude on rainfall. A comparison of probabilities of exceedence of maximum rainfall intensities for each sub-catchment yielded no clear difference in rainfall intensities between the catchments (Rijsdijk and Bruijnzeel 1990).

The differences in water balances (Table 5.5) and runoff coefficients (Table 5.6) between the three sub-catchments are remarkable, as watertight (forested) catchments under similar climatic and geological circumstances typically exhibit runoff ratios of about 50% (Grobbe 1989). As such, it is clear that the Manting sub-basin receives large amounts of external groundwater flow whilst the

Coban Rondo sub-basin loses substantial amounts via underground leakage. The Sayang sub-basin, on the other hand, does not seem to gain or lose water.

This situation could be explained by the geological and topographic situation of the Konto river basin (Fig. 5.1). The Coban Rondo sub-catchment is situated on the outer rim of a collapsed volcanic cone, while the Manting basin is part of the inner cone of a volcano. Since, in both cases, the slope of the inner cone is much steeper and longer than the slope on the outside, flow from the outer rim towards the inner cone is plausible. A similar case has been described in Purwanto (1999). In the Coban Rondo sub-catchment, the measured volumes of areal runoff were actually higher in the upper part than in the lower part. This can probably be explained by water use for irrigation in the lower part or leakage through the highly permeable debris in the lower part of the basin, a phenomenon which also has been described by Meijerink (1976) in the Serayu river basin (Java, Indonesia).

5.5.2 Quickflow

The variations in quickflow (Table 5.6) illustrate the different hydrological behaviour of the upper and lower parts of the three sub-catchments Since land use in the upper parts of the three sub-basins is comparable (see Table 5.1), factors other than land use should account for the differences in quickflow volumes. It is evident that the high infiltration capacity of the deep and permeable soils in the natural forest, shrubland and plantation forest of the upper basins (Rijsdijk and Bruijnzeel 1990) virtually impedes any Horton overland flow (HOF), even in response to an extreme extent (Bruijnzeel 1983). Only in the upper Sayang, which contains some rain-fed agriculture (4 ha) and a few trails in agroforestry fields (2.6 ha), could a limited amount of HOF be generated. In addition to this, direct channel precipitation would also contribute to quickflow. This, however, would be limited in view of the small area occupied by the channels (upper Coban Rondo, maximum 2 ha and upper Manting and Sayang, maximum 1 ha). Other mechanisms such as saturated overland flow (SOF) have never been observed on a large scale in the upper parts of the catchments (Rijsdijk and Bruijnzeel 1990).

More likely could be the generation of stormflow in the upper basins through subsurface stormflow (SSSF), which is common on well-drained soils on deep slopes (Dunne 1978). Such stormflow could eventually be boosted by the push-through mechanism (Hewlett and Hibbert 1967), or by pipe flow, which is possible in humid areas with a long rainy season (Hadley et al. 1985; Bruijnzeel 1983). Additionally, the generation of temporary groundwater ridges might play a role (Ward 1984a). This mechanism could well explain the higher relative volume of quickflow and maximum daily flows found in the Manting upper basin[8] than in the

---

[8] When the quickflow is expressed as fraction of streamflow the value for Manting might have been somewhat supressed by the relatively high groundwater contribution (Sect. 5.1).

Sayang and, to an even more pronounced extent, the Coban Rondo basin. The respective gain and leakage in Manting and Coban Rondo, from and to catchments out of the project area, will raise and lower the groundwater table in these sub-catchments, and, consequently, will raise and lower the volume of the quickflow, respectively. The hydrology of the lower part of the sub-catchments is strongly related to land use. At the lower stations at Coban Rondo (CB) and Sayang (SB), which have many impermeable surfaces such as settlements, roads and also inundated rice fields (Table 5.1), the amount of quickflow is double that observed in the upper parts of these catchments. By contrast, in lower Manting, where there are virtually no impermeable surfaces, quickflow hardly increased relative to that in upper Manting. The influence of development on quickflow is clearly showing in the study of Ismail (1997). An undisturbed catchment had a mean monthly quickflow contribution of 15% in contrast to a nearby developed catchment with a quickflow contribution of 22% (Table 5.7) The same table makes also visible the influence of the soil on quickflow, forested watersheds with shallow or clayey soils show much higher quickflow contributions (22–51%) than watersheds with deep volcanic soils (4–22%). The vegetation at the Skudai river catchment was oil palm, but according to Chong et al. (2006), the response of oil palm catchment did not differ much from forest.

The nature of the runoff can also be illustrated with the use of the minimum contributing area (MCA) concept (Dickinson and Whiteley 1970). The MCA has been defined as the minimum area which, contributing 100% of the effective rainfall, would yield the measured storm runoff. When comparing the estimated surface area of the locations that may possibly contribute HOF and SOF (e.g. channels, settlements and inundated rice fields; Table 5.1), with the calculated MCA values (Table 5.6), it is clear that the relationship between quickflow and HOF plus SOF is weak.

For the upper stations, the calculated MCAs were larger than the areas which potentially could produce HOF and SOF. This may be explained by the above-mentioned contribution of SSSF. For the lower stations of Sayang and Coban Rondo, the calculated MCAs were smaller than the areas producing HOF and SOF. This could probably be explained by the hydraulic behaviour of rice fields; these delay quickflow during low-rainfall events and enhance quickflow during high-rainfall events (Godefroy 1931; Purwanto 1999). Other factors are the fact that rice fields are not always inundated and some runoff from settlements is often diverted to agricultural fields instead of running straight to the rivers (Rijsdijk, personal observation).

5.5.3 Sediment Yields

The suspended sediment yield for the sub-catchments shows large variations, both between catchments as well as between the upper and lower stations (Table 5.8). In particular, the high suspended sediment yield in the upper Sayang catchment

(SA) is remarkable when compared with that in the other upper catchments (average values of suspended sediment in Mg ha^{-1} for the hydrological years 1988 to 1989: CA 0.35; MA 0.95; and SA 3.4). Causes such as differences in quickflow (Table 5.6) might partly be responsible for the differences in suspended sediment between CA and SA, but cannot explain the difference between the MA and the SA station. Lower soil stability or higher rainfall erosivity (Rijsdijk and Bruijnzeel 1990) can be ruled out.[9] Some of the load is likely to have been contributed by plantation forest, agroforestry fields, and trails (Rijsdijk 2005; Rijsdijk et al. 2007a, b). However, from the equally high amount of bedload (0.46 Mg ha^{-1} or 11.9% of the total load at SA), it was estimated (Sect. 4.2) that around 1.4 Mg ha^{-1} or 36% of the total sediment load originates from (riparian) mass wasting and bank or channel erosion. The upper Coban Rondo and Manting streams in particular carried very little bedload (CA 1.8 vs. MA 0.5% of the total load). This reflects the resistance to erosion of their riparian zones (consisting of lava flows and andesite boulders, respectively), in contrast to the less-resistant lahar flows found in the upper Sayang sub-catchment.

The much higher suspended sediment yields of the lower parts of the Coban Rondo (CB) and Sayang sub-catchment (SB) (9.27 vs. 11.15 Mg ha^{-1}) reflect the influence of other sediment sources such as unpaved roads, dry agricultural land, and settlements (Table 5.1, Rijsdijk et al. 2007a; Rijsdijk 2005). This is in contrast to low values at MB (0.95 Mg ha^{-1}), which has no settlements or agricultural activities in the lower part. The somewhat higher sediment yield of SB as compared to CB can be explained by the higher percentages of quickflow (Table 5.6), or by hidden sediment sources (landslides or scour from channels draining the rice terraces) as the surface erosion in the Coban Rondo is actually higher than in Sayang (average CB 13.3 vs. SB 9.6 Mg ha^{-1}, see Rijsdijk 2005). The amount of bedload is comparable for the two sub-catchments (both about 6.5% of the total load). The higher amount of bedload at MB as compared to MA is probably the result of damage to the riverbank caused by the clearing of the shrubs along the river.

The high sediment yield at the MB station in 1990, at least 5.5[10] Mg ha^{-1} or more than five times the annual average, were caused by a number of extreme events. In January 1990 several storms amounting to 1,594 mm or nearly half of the average annual rainfall, resulted in at least 770 (see footnote 10) mm outflow. This high sediment yield, caused by an extreme event is not unusual, as other studies in the region demonstrate. For example, Loughran (1989), showed that in a partially cultivated catchment in Australia, 79% of 3 year sediment yield was transported by only three events.

As described in (Rijsdijk 2005), the average sediment yields of the three sub-catchments are consistent with the estimated siltation rates of Lake Selorejo

---

[9] Experiments using Wischmeier and Smith (1978) erosion plots showed that soil losses were not consistently higher in Sayang than in the other sub-catchments (Rijsdijk and Bruijnzeel 1990).
[10] Probably underestimated due to damage to the equipment.

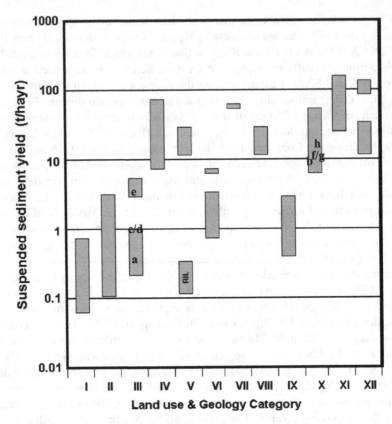

Fig. 5.3 Ranges in reported catchment sediment yields in southeast Asia as a function of geological substrate and land use (adapted from Bruijnzeel 2004). Categories: *I*, forest, granite; *II*, forest, sandstones/shales; *III, forest, volcanics*; *IV*, forest, marls; *V*, logged (RIL: reduced impact logging); *VI*, cleared, sedimentary rocks (*lower bar*: micro-catchments); *VII*, cleared, volcanics; *VIII*, cleared, marls; *IX*, medium-large basins, *mixed land use*, granite; *X*, *idem, volcanics*; *XI*, idem, volcanics plus marls; *XII*, urbanised (*lower bar*), mining and road building (*upper bar*). Catchment sediment yields from this study (in Mg ha^{-1}): a = sediment yield at station CA (0.4), b = sediment yield at station CB (9.9), c = sediment yield at station MA (1.0), d = sediment yield at station MB (1.1), e = sediment yield at station SA (3.9), f = sediment yield at station SB (11.7), g = sediment yield of the Konto river catchment (low estimate*) (11), h = sediment yield of the Konto river catchment (high estimate*) (16)

(11–16 Mg ha^{-1}) (Fisch 1983; Brantas 1989), taking into account the sediment delivery ratio, the contribution of an additional catchment containing many gullies, and the influence of the high sediment yield of the Konto river[11] [22 Mg ha^{-1} during the very wet (3,433 mm) year of 1990 when many landslides occurred (Rijsdijk et al. 2007b)]. In Fig. 5.3 (from Bruijnzeel 2004) the sediment yields of this study have been compared with the results from more than 60 studies of

---

[11] Measured at Kambal station, Fig. 5.1.

catchment sediment yields in south-east Asia as a function of geological substrate, landcover and the degree of disturbance (Bruijnzeel 2004). From this comparison one can conclude that all values from this study are within the range of comparable studies.

As expected in these narrow valleys the possibilities for sediment storage are low and hence the sediment delivery ratios (SDR) are rather high (Rijsdijk 2005). The highest values occur in the upper stations CA (1.1), MA (2.4) and SA (1.1) where the possibility for storage in the deeply incised valleys is virtually zero. The SDRs of the lower stations CB (0.7) MB (0.7) which have more space for sediment deposition are in line with the values (0.4–0.07) reported by Purwanto (1999) for a 105 ha catchment. The relatively high SDR of the Sayang lower station SB (1.2) might be caused by quick removal of sediment by the relatively high amount of quickflow, or by overlooked sediment sources such as landslides and channels from the rice fields (see Rijsdijk 2005).

5.5.4 Influence of Different Calculation Methods on Sediment Yields

In Indonesia, river sediment is usually sampled during office hours, while daily streamflow is computed by inserting the average water level of the day into the stage discharge relationship (Anonymous 1989). This average discharge is then inserted into the sediment-rating curve to obtain the average sediment discharge for the day. This methodology could cause serious underestimation of the sediment yield for two reasons.

First, sampling only during office hours will miss most high floods as, in general, they occur late in the afternoon or in the evening. This seriously underestimates the sediment load, as the bulk of sediment transport takes place during high flows (Turvey 1974; Amphlett 1988; Rijsdijk and Bruijnzeel 1990). Correct procedures should include both the rising stage and the falling stage, as sediment curves are often skewed (Williams 1989; Rijsdijk and Bruijnzeel 1990). To evaluate the effect of this practice and to be able to compare the project results with sediment yields computed for other catchments in Java, the sediment data that were collected between December 1988 and February 1989 at the upper Sayang[12] station were processed in two ways. Sediment rating curves were derived for the entire data set and for the samples taken during office hours only. The result was an underestimation of suspended sediment yield of about 33% (0.76 vs. 1.13 Mg ha^{-1}) due to sampling in office hours only.

---

[12] From 1 November 1988 to 15 February 1989 comprising 1150 mm rain (50% of the total rainfall in 1988–1989).

The second reason that sediment yields could have been underestimated is the use of average discharge values for the sediment rating curve. This is likely to result in a large underestimate of the true mean, depending on the peak of that day's hydrograph, as both discharge and sediment rating curves are power curves. Re-calculation of the sediment yield using the above-mentioned data in the conventional way resulted in a yield of 0.65 Mg ha^{-1}. This implies an underestimate of 47% overall, when compared to the sediment yield (1.20 Mg ha^{-1}) calculated using the event-based method with the full data set (details in Kaatee 1989).

5.6 Conclusions

The hydrological behaviour of the three upper catchments can mainly be explained by the geomorphology and, to a minor extent, also by land use. The most important contributor to quickflow is the subsurface stormflow, although Horton overland flow also played a role in the Sayang upper catchment containing some agroforestry. The hydrology of the lower part of the sub-catchments is strongly related to land use, where the influence of impermeable areas like settlements and roads, and also the specific hydrologic behaviour of inundated rice fields, is clearly visible.

The amount of suspended load ranges from 0.35 Mg ha^{-1} in upper Coban Rondo up to 11.15 Mg ha^{-1} in lower Sayang. The relative amount of bedload in the rivers ranges from nearly 0.5% (MA) to about 12% (SA). This is the equivalent of 1–36% of the total load (estimated values). This relatively high volume of bedload in the Sayang upper catchment can mainly be attributed to the less resistant riverbanks of the lahar zone. These results stress the need to include the bedload component in assessments of the sediment yield of rivers.

Both the amount of suspended load and the bedload in the lower catchments are clearly related to land use (Table 5.1). The amount of bedload for the two catchments is comparable (about 6.5%), but the amount of suspended load for the Sayang lower station is somewhat higher; this was probably caused by unmeasured sediment sources or more efficient transport due to higher amount of quickflow, and not due to differences in land use (Rijsdijk 2005).

The official data on sediment yields in rivers in Indonesia are likely to be seriously underestimated due to incorrect sampling procedures and calculation methods. In addition, the amount of bedload is seldom measured and it is likely that the sediment levels of extreme events are often underestimated.

Acknowledgments The author gratefully acknowledges the support of his colleagues within the Konto river project, notably the team leaders Sjaak Beerens, M.Sc. (DHV Consultants) and Bapak Bambang Moerdiyono, M.Sc. (DGRRL); Dr. Sampurno Bruijnzeel (Free University) (scientific supervision); Edi Hertanto, Jumali, Suparno, and the M.Sc. students Evert Kaatee, Chris Bremmer and Theo Prins (data collection and processing, laboratory analyses); Ahmad Zaeni (supervision of construction work); and the numerous temporary field assistants that helped out during the intensive campaigns that generated the results described in this paper. The Konto river project was financed by the Directorate General of International Co-operation (DGIS) of the Netherlands under project no. ATA 206.

References

Ambar S, Wiersum KF (1980) Comparison of different erodibility indices under various soil and land use conditions in West Java. Indonesian J Geogr 39:1–15

Amphlett MB (1988) A nested catchment approach to sediment yield monitoring in the Magat catchment, Central Luzon, the Philippines. In: Proceedings of the 5th international soil conservation conference, Bangkok, Thailand, 18–29 January 1988, pp 16

Anonymous (1989) Konto river basin project ATA 206 inception report. Konto river project communication no. 1, Konto river project, Malang, Indonesia

Balamurugan G (1991) Sediment balance and delivery in a humid tropical urban river basin: the Kelang river, Malaysia. Catena 18:271–287

Bogen J (1992) Monitoring grain size of suspended sediments in rivers. In: Erosion and sediment transport programmes in river basins. IAHS publication 210, pp 183–190

Brabben TE (1979) Reservoir sedimentation study, Selorejo, East Java, Indonesia. Reservoir survey and field data. Hydraulics research station report OD 15. HRS, Wallingford

Brantas PU (1989) Laboran Pengukuran Sedimentasi Waduk Selorejo. Unpublished report (in Indonesian). Public Works (P.U.). Department Directorate of Irrigation, Malang, Indonesia

Bruijnzeel LA (1982) Hydrological and biologichemical aspects of man-made forest in south central Java, Indonesia. Ph.D. Thesis, Vrije Universiteit, Amsterdam, The Netherlands

Bruijnzeel LA (1983) Evaluation of runoff sources in a forested basin in a wet monsoonal environment: a combined hydrological and hydrochemical approach. IAHS Publ 140:165–174

Bruijnzeel LA (1988) Forest hydrology in Indonesia: challenges and opportunities. Konto river project communication no. 7, Konto river project, Malang, Indonesia

Bruijnzeel LA (2004) Hydrological functions of tropical forests: not seeing the soil from the trees. Agric Ecosyst Environ 104:185–228

Chong MH, Zulkifli Y, Ayob K (2006) Rainfall-runoff characteristics in a small oil palm catchment. In: Proceedings of water for sustainable development towards a developed nation by 2020, Guoman resort port dickson, Malaysia, 13–14 July 2006

Dickinson WT, Whiteley H (1970) Watershed areas contributing to runoff. IAHS publ 96:12–26

Diemont WH, Mannaerts C, Nurdin, Smiet AC, Rijnberg T (1991) Re-thinking erosion on Java. Paper prepared for the international workshop on conservation farming policies of hillslopes for the development and sustainability of natural resources, Solo, Indonesia, 8–15 March 1991

Dunne T (1978) Field studies of hillslope flow processes. In: Kirkby MJ (ed) Hillslope hydrology. Wiley, Chichester, pp 227–294

Emmett WW (1981) Measurement of bedload in rivers. In: Erosion and sediment transport measurements. IAHS publication 133, pp 3–16

FAO (1990) Soil map of the world. Revised legend. Reprinted with corrections. World resources report no. 60. Food and Agriculture Organization of the United Nations, Rome

Ferguson RI (1986) River loads underestimated by rating curves. Water Resour Res 22(1):74–76

Fisch IL (1983) Reservoir sedimentation study, Selorejo, East Java, Indonesia, Reservoir survey June 1982. Hydraulics research station (HRS) report, OD 51. HRS, Wallingford, UK

Godefroy PW (1931) De invloed van geinudeerde sawahs op den maximum afvoer van een stroomgebied. De waterstaats ingenieur May 1931, pp 189–194 (in Dutch)

Grobbe HW (1989) The water balance of a thirty year old Pinus merkusii plantation forest in upland West Java, Indonesia. M.Sc. Thesis, Free University, Amsterdam

Guy HP, Norman VN (1970) Field methods for measurements of fluvial sediment. Techniques of water resources investigations of the US geological survey, Book 3 Chapter C2. US government printing office, Washington DC

Hadley RF, Lal R, Onstad CA, Walling DE, Yair A (1985) Recent developments in erosion and sediment yield studies. International Hydrological Programme, UNESCO, Paris

Hewlett JD, Hibbert AR (1967) Factors affecting the responses of small watersheds to precipitation in humid areas. In: Sopper WE, Lull HW (eds) International symposium on forest hydrology. Pergamon Press, Oxford, pp 275–290

Ismail WR (1997) The impact of hill land clearance and urbanization on runoff and sediment yield of small catchments in Pulau Pinang, Malaysia. IAHS publication 245

Jahani A (1992) Calculating the suspended sediment load of the Dez river. IAHS Publ 210:219–224

Kaatee EG (1989) Short-term rainy season sediment yield in the Sayang catchment, Kali Konto upper watershed, East Java, Indonesia. M.Sc. Thesis, Free University, Amsterdam

Loughran RJ (1989) The measurement of soil erosion. Prog Phys Geogr 13(2):216–233

Meijerink AMJ (1976) On the nature of baseflow and groundwater occurrences in the Serayu river basin. In: Seraya valley project (Java Indonesia) final report, vol 2, 55–64

Miller CR (1951) Analysis of the flow duration, sediment rating curve method of computing sediment yield. US bureau of reclamation report

Nuffic-Unibraw (1984) Soils and Soil Conditions. Kali Konto upper watershed, East Java. Main Report. Nuffic-Unibraw Soil Science Project. Universitas Brawijaya, Malang

Oldeman LR (1975) An agro-climatic map of Java. Contributions of the central research institute for agriculture, Bogor

Østrom G (1975) Sediment transport in glacial meltwater streams. In: Jopling A, McDonald BG (eds) Glaciofluvial and glaciolacustrine sedimentation. Society of economic paleontologists and mineralogists special publication no 23, 101–122

Purwanto E 1999. Erosion, sediment delivery and soil conservation in an upland agricultural catchment in West Java, Indonesia. Ph.D. Thesis, Vrije Universiteit, Amsterdam, The Netherlands

Rijsdijk A (2005) Evaluating sediment sources and delivery in a tropical volcanic watershed. In: Sediment budgets 2, IAHS publication 292, pp 16–23

Rijsdijk A, Bruijnzeel LA (1990) Erosion sediment yield and land-use pattern in the Upper Konto watershed, East Java, Indonesia. Konto river project communication no. 18 Vols I, II, DHV Consultants, Amersfoort

Rijsdijk A, Bruijnzeel LA (1991) Erosion sediment yield and land-use pattern in the Upper Konto watershed, East Java, Indonesia. Konto river project communication no. 18, vol III, Results of the 1989–1990 measuring campaign. DHV Consultants, Amersfoort, The Netherlands

Rijsdijk A, Bruijnzeel LA, Sutoto KC (2007a) Runoff and sediment production from rural roads, trails and settlements in the Upper Konto catchment, East Java, Indonesia. Geomorphology 87:28–37

Rijsdijk A, Bruijnzeel LA, Prins TM (2007b) Sediment yield from gullies, riparian mass wasting and bank erosion in the Upper Konto catchment, East Java, Indonesia. Geomorphology 87:38–52

RIN (1985) Evaluation of forest land. Kali Konto upper watershed, East Java, vol III, Natural forest. Research institute for nature management, Leersum

Rutten LMR (1917) On the rate of denudation on Java. Verslagen van de Koninklijke Nederlandsche Academie van wetenschappen, Amsterdam 1917, pp 920–930

Thomas RB (1988) Monitoring baseline suspended sediment in forested basins: the effect of sampling on suspended sediment rating curves. J Hydrol Sci 33(5):499–514

Thorne CR (1981) Field measurements of rates of bank erosion and bank material strength. In: Erosion and sediment transport measurement, IAHS publication 133, Louisiana, pp 503–512

Turkelboom F (1999) On-farm diagnosis of steepland erosion in northern Thailand. Ph.D. Thesis, Catholic University of Leuven, Leuven, Belgium

Turvey ND (1974) Nutrient cycling under tropical rain forest in Central Papua. University of Papua New Guinea, Department of geography occasional paper no. 10

Van der Linden P (1978) Contemporary soil erosion in the Sanggeman river basin related to the quaternary landscape development. Ph.D. Thesis, University of Amsterdam

Van Dijk JW, Ehrencron VKR (1949) The different rate of erosion within two adjacent basins in Java. Communications of the general agricultural experiment station. Bogor 84:1–10

Walling DE (1977a) Assessing the accuracy of suspended sediment rating curves for small basins. Water Resour Res 13:531–538

Walling DE (1977b) Limitations of the rating curve technique for estimating suspended sediment loads, with particular reference to British rivers. IAHS Publ 122:34–48

· Walling DE, Webb BW (1988) The reliability of rating curve estimates of suspended sediment yield: some further comments. IAHS Publ 174:327–337

Walling DE, Webb BW, Woodward JC (1992) Some sampling considerations in the design of effective strategies for monitoring sediment-associated transport. In: Erosion and sediment transport monitoring programmes in small rivers, IAHS publication 210, pp 279–288

Ward RC (1984a) On the response to precipitation of headwater streams in humid areas. J Hydrol 74:171–189

Ward PRB (1984b) Measurement of sediment yields. In: Hadley RF, Walling DE (eds) Erosion and sediment yields, some methods of measuring and modelling. Geobooks, Norwich

White SM (1988) Catchment scale monitoring of erosion and sediment yield. World Bank Colloqium, Washington

Williams GP (1989) Sediment concentration versus water discharge during single hydrologic events in rivers. J Hydrol 111:89–106

Wischmeier WH, Smith DD (1978) Predicting rainfall erosion losses agricultural research service handbook no. 282, US Department of Agriculture

Chapter 6
Ecological Protection and Restoration in Sanjiangyuan National Nature Reserve, Qinghai Province, China

Xi-lai Li, Gary Brierley, De-jun Shi, Yong-li Xie and Hai-qun Sun

Abstract Historical data and published results are reviewed to assess the rationale and design criteria used to establish the Sanjiangyuan (Three Source Region) National Nature Reserve in Qinghai Province, China. This area, which comprises the headwaters of the Yellow, Yangtze and Mekong Rivers, has been described as the 'Third Pole' or the 'Roof of the Earth'. Key drivers for the designation of this ecological reserve include concerns for sustainable water resources management and biodiversity management, especially the protection of endangered flora and fauna. These issues are placed in context of prevailing and prospective threats, namely climate change, environmental degradation associated with overgrazing and burrowing mammals, and human-induced pressures (population growth and various land use practices). Ecological protection and restoration measures being applied in the Sanjiangyuan region are reviewed. The approach to environmental conservation and management parallels initiatives applied in many other parts of the world, with 18 core areas, connected and/or surrounded by buffer areas, with experimental areas beyond. Adopted measures are framed primarily in relation to wetland functions, striving to protect water resources and the precious but fragile ecosystems in this region. Water resources planning strategies and landscape ecology programmes link lake and river ecosystems to grassland, forest and wetland management strategies at the landscape scale.

X. Li · Y. Xie · H. Sun
Agricultural and Animal Husbandry College, Qinghai University, Xining, China

G. Brierley (✉)
School of Environment, University of Auckland, Private Bag 92019,
Auckland, New Zealand
e-mail: g.brierley@auckland.ac.nz

D. Shi
Qinghai Agriculture and Animal Husbandry Bureau, Xining, China

D. Higgitt (ed.), *Perspectives on Environmental Management and Technology in Asian River Basins*, SpringerBriefs in Geography,
DOI: 10.1007/978-94-007-2330-6_6, © The Author(s) 2012

Keywords Conservation · Rehabilitation · Ecological reserve · Grassland · Mountain region · Ecosystem management

The preservation of the full range of biodiversity and of physical features is an essential element in the selection of mountain protected areas. As an integral part of planning, provision should be made for the protection of large examples of natural ecosystems and of populations of plant and animal species, together with sites illustrating the principal geological and physiographic features and the processes at work in the landscape. These should be supplemented by the protection of a larger number of small areas representing the full local variety of species and ecosystems, including intra-specific genotypic variation (Hamilton and Macmillan 2004, p. 14).

6.1 Introduction

Ecosystems provide essential goods and services, contributing substantially to social and economic development. Societal wellbeing is dependent upon the maintenance of biodiversity and the capacity of species and ecosystems to survive, evolve and adapt to environmental change. The extent and impact of human disturbance upon 'natural' ecosystems across the world is no longer in dispute (Millennium Ecosystem Assessment 2005). Environmental damage brought about by population pressure, land use intensification and climate change, among many factors, has been accompanied by progressive depletion of natural resources. In response to these concerns, genuine commitment to environmental protection has become a core element of sustainable management programmes. Although protected areas now cover over 12% of the Earth's land surface (Chape et al. 2005), conservation activities in their own right are unable to maintain the inherent diversity of genes and ecosystems. An ecosystem approach to environmental management promotes recovery and enhancement of the self-sustaining properties of 'natural' systems, recognizing implicitly that there are too many species to save them one at a time.

Because of their history, isolation, and great variability of habitat, mountain regions are treasures of high biodiversity and are rich in endemic species (Hamilton and Macmillan 2004). These reservoirs of biodiversity contain rich assemblages of species (and their genotypes). Distinctive ecosystems have developed in response to the pronounced altitudinal belts and local variability in physical conditions in these areas. However, these steep, high altitude terrains are especially susceptible to human-induced damage, and many elements of these fragile environments are vulnerable to accelerated rates of climate change.

In global terms, concerns for environmental conditions in the face of population pressures and the demands for economic development are nowhere more evident than in China. In the 11th Five Year Plan, explicit recognition is given to the need to enhance and expand approaches to environmental protection (Xiao 2006). One of the most important initiatives is efforts to protect the headwater zone of the Yellow, Yangtze and Mekong Rivers in the high altitude country of the Qinghai-Tibetan Plateau (see Fig. 6.1). This area in Qinghai Province is known as the

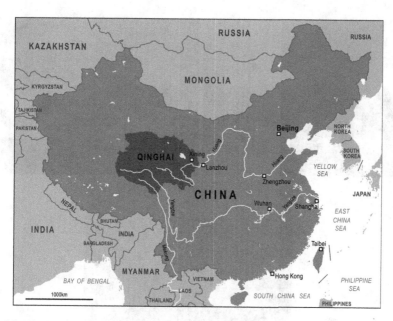

Fig. 6.1 Qinghai province, western China. The headwater zone of the Yellow, Yangtze and Mekong rivers lies in the high altitude country of the Qinghai-Tibetan Plateau

Sanjiangyuan (the source zone of the Three Rivers), and is often referred to as 'The Third Pole' or 'The Roof of the World'. The Sanjiangyuan region is the highest (average elevation above 4000 m) and largest wetland ecosystem in the world (Li 2007). Indeed, this region is also known as the 'kidney of the earth', the 'cradle of living forms' and 'the water tower of China'. Key functions of these fragile and complex wetland ecosystems include their capacity to retain water, reduce surface runoff, store flood-water to mitigate flood events, reduce pollution, maintain biodiversity, supply water for human services, regulate climate, etc. Annual water supply from these three rivers to downstream areas is as much as 60 billion m^3, supplying 49% of total water to the Yellow River, 25% to the Yangtze and 15% to the Mekong (Li 2007; Ma 2007; Zhou and Yie 2007). Although the drainage areas of the Yellow and Yangtze Rivers make up 24% of the total land area of the country, they comprise 50% of the total population of the country and make up 50% of the total Gross Domestic Product (Chen et al. 2002). Protection of water resources is integral to China's economic and environmental security (see Cannon 2006).

The high elevation and dry, cold climate of the Qinghai-Tibetan Plateau foster the unique fauna and flora (alpine germplasm) of the Sanjiangyuan region (see Fig. 6.2). Spatial distributions of many living forms have been marginalized in this area. The fragility of ecosystems is locally threatened by increases in human population and uncontrolled production activities (Ma et al. 2000; Zhou et al. 2003b; Du et al. 2004; cf., Harris 2010). Removal of natural vegetation has reduced the water-retention capacity of the region and lowered the capacity of the environment to cope and recover in periods of stress (Li 2007). Natural disasters such as floods, droughts and

Fig. 6.2 Distinctive topographic, faunal and floral attributes of the Sanjiangyuan region

dust storms occur more frequently, and erosion has become more serious. Overgrazing has caused the degradation of grassland, seriously affecting living condition of herders (Ma et al. 2001, 2002; Zhou et al. 2003a, b; Wang et al. 2004; Shi et al. 2005). Desertification is growing as uncontrolled harvesting of wood, gathering of materials for herbal medicines (especially caterpillar fungus (*Cordyceps sinensis*); see Winkler 2009)) and mining activities degrade grassland and forest vegetation (Chen et al. 2002). Deterioration and fragmentation of habitat is decreasing the regional biodiversity (Chen et al. 2007).

Implementing the policy of 'the Great Development of the West' is an important strategy laid out by the Chinese Government (see Economy 1999, Goodman 2004, Xin 2008). An inspection team was sent by the 'Jiu San Society' to examine the eco-environment of the Sanjiangyuan region (Chen et al. 2007). With support from

Fig. 6.3 The Sanjiangyuan region makes up the southern half of Qinghai province in western China. The region is divided into 17 administrative districts

relevant departments of central government, the People's Government of Qinghai province has prepared a programme for ecology-based protection and construction in the area, including the formation of the Sanjiangyuan National Nature Reserve. A comprehensive, integrated and balanced approach to protection and construction strives to ensure that economic development continuously enhances the living standards of local people (Chen et al. 2007). To achieve this, the distribution and livelihood of human beings is being adjusted, encouraging herders to develop more advanced approaches to livestock production within an ecology-oriented economy. Alternatively, many are being asked to abandon livestock production.

In this manuscript we review the natural features and ecological values of the Sanjiangyuan region, appraise the pressures and stressors that bring about ecological degradation, and highlight measures that are being taken to address these issues through the creation of the Sanjiangyuan National Nature Reserve. The design criteria applied in the derivation of the reserve and suggested measures for ecological protection and restoration are discussed in relation to large scale conservation initiatives elsewhere in the world.

6.2 Regional Features and Ecological Conditions of the Sanjiangyuan Region

The Sanjiangyuan region lies at the heart of the Qinghai-Tibetan Plateau in the southern part of Qinghai province (latitude 31°39′–36°12′ N, longitude 89°45′–102°23′ E) (see Figs. 6.1 and 6.3). It covers a total area of 363,000 km²,

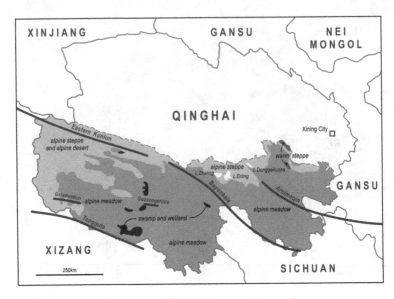

Fig. 6.4 Vegetation map of Qinghai province (Liu et al. 2004)

accounting for 50.4% of the total area of Qinghai province. The major mountainous areas such as the eastern Kunlun, Arnimaqin, Bayankala and Tanggula
Mountains range in elevation from 3335 to 6564 m, typically between 4000 and
5800 m (see Fig. 6.4; Sun and Zheng 1996; Chen et al. 2002). Snow-covered
mountains cover a total area of 2,400 km^2, with 1,812 km^2 covered by continental
mountain glaciers. Glaciers are retreating at a rate of 30–50 m/year (Zheng 1996;
Chen et al. 2002). Wetlands account for 20.3% of the total land area (73,300 km^2)
(Zhou and Yie 2007). The Sanjiangyuan region has more than 16,500 lakes, of
which around 190 have an area > 0.5 km^2, and there are extensive groundwater
reserves (Chen et al. 2002; Zhou and Yie 2007). Numerous lakes and swamps are
included on the List of Important Wetlands in China.

The alpine continental climate is characterized by clear separation of dry and wet
seasons but small annual temperature differences, large diurnal and seasonal temperature differences, long sunshine hours and strong radiation (Zhang et al. 1999).
Mean annual temperature across the region ranges from −5.6 to 3.8°C while mean
annual precipitation ranges from 262 to 773 mm. The cold season, which lasts for
7 months, is controlled by high pressure systems. Given the high altitude and very
thin air, the growing period is short and there is no absolute frost-free period.

Primary vegetation classes in the region include coniferous forest, broad-leaf
forest, needle-leaf and broad-leaf mixed forest, shrub, meadow, steppe, swamp and
aquatic vegetation, cushion plants, and zones of sparse vegetation (Table 6.1).
Forests comprise 1.22 million ha in the Sanjiangyuan region (3.27% of the land
area). Grasslands account for 59% of the total land area (21.4 million ha, of which
19.27 million ha is utilizable grassland). Alpine meadow (82.6%) and alpine
steppe (12.1%) are the main grassland types (see Fig. 6.4). Alpine meadow is

Table 6.1 Land utilization in the Sanjiangyuan region (Statistical Bureau of Qinghai Province 2003)

Prefecture	Total land area (10⁴ km²)	Grassland		Woodland area (10⁴ ha)	Of total woodland		Tilled and ploughed land (10⁴ ha)	Water area (10⁴ ha)	Others (10⁴ ha)
		Grassland area (10⁴ ha)	Acceptable grassland area (10⁴ ha)		Wood area (10⁴ ha)	Shrub forest area (10⁴ ha)			
Total	36.31	2141.80	1926.60	122.40	11.33	107.40	4.81	106.34	1255.65
Yushu (all counties)	19.79	1084.80	956.98	41.40	4.67	35.33	1.64	79.05	
Guoluo (all counties)	7.63	675.40	625.52	57.47	3.60	53.20	0.29	25.32	
Hainan (only Xinghai and Tongde county)	1.69	145.73	139.16	14.13	2.00	11.13	2.44	0.86	
Huangnan (only Zeku and Henan county)	1.31	125.69	119.92	9.40	1.07	7.73	0.44	1.11	
Haixi (only Tanggula town)	5.87	110.19	110.19						

Note Qumalai, Zhidou, Zhadou, Nangqian, Yushu and Chengdou counties are in Yushu prefecture and Madou, Dari, Gande, Maqin, Jiuzhi and Banma counties are in Guoluo prefecture mainly distributed on mountain slopes, or on top of mountains, terraces, or valley benches at elevations between 3200 and 4700 m. The vegetation comprises cold-tolerant perennial plants. The main grassland species include sedge (*Carex* spp., especially *Kobresia* spp.), needlegrass (*Stipa* spp.), saussure (*Saussurea* spp.), roegneria (*Roegneria* spp.), bluegrass (*Poa* spp.), wild ryegrass (*Elymus* spp.) and speargrass (*Achnatherum* spp.) (see Fig. 6.2c). The height of forage is low, ranging between 5 and 30 cm. Productivity is very low. Around 12.47 million ha (58.3% of the total grassland area) has been subjected to moderate to severe degradation (Chen et al. 2002; Wang et al. 2005; cf., Harris 2010; Li et al. 2011).

Fauna in the Sanjiangyuan region comprise 93 species of mammals belonging to 22 families of 8 classes; 255 species of birds belonging to 41 families of 16 classes; and 18 species of amphibians and reptiles belonging to 10 families (Chen et al. 2007). Among the mammals and birds, 69 species are on the list of "protected wildlife of national importance", of which 16 animals are 'first class' protected species (e.g. Tibetan antelope (*Pantholops hodgsoni*), wild yak (*Bos grunniens*) and snow leopard (*Uncia uncia*)) and 53 animals are 'second class' protected species (e.g. blue sheep (*Pseudois nayaur*) and Tibetan gazelle (*Procapra picticaudata*)).

The Sanjiangyuan region is a vast area with very low population density. Of the total regional population of 590,000, 408,900 are herders (69.3%) (Table 6.2). Tibetan families account for 90% of the herding families, with small number of other nationalities such as Mongolian. Livestock grazing is the main component of the regional economy, with over 22 million sheep in the region. Since 1990, herders have rented land from the local government over 50–100 year lease periods. Under this arrangement, herders can use and manage the land tax free under laws such as the grassland law and the land utilization law, in programmes co-ordinated by the grassland supervision station.

6.3 Key Mechanisms and Stressors that Bring About Environmental Damage

Establishment of the Sanjiangyuan National Nature Reserve is a coordinated response to many social and environmental pressures in this fragile region. Prior to assessing criteria used in the design and implementation of the ecological reserve, the primary stressors that induce environmental damage in the region are outlined.

6.3.1 Climate Change

Global climate change is the fundamental natural factor causing the deterioration of ecological environments in the Sanjiangyuan region (Zhang et al. 1998, 1999). Fragile ecosystems have been destabilized, reducing their capacity for self-restoration.

Table 6.2 Socio-economic statistics for the Sanjiangyuan region (Statistical Bureau of Qinghai Province 2003)

Prefecture	Total population (10⁴ persons)	Population of pastoral area (10⁴ persons)	Households of pastoral area	Annual income per capita (Yuan)	Livestock number (10⁴ Cattle/sheep)	Equivalent sheep units (10⁴ sheep units)
Total	59	40.89	83,531	1549.69	1038.93	2224.03
Yushu(all counties)	26.3	17.04	34,664	1432.5	278	748.81
Guoluo(all counties)	13.54	9.20	20,732	1588.97	232.2	759.90
Hainan (only Xinghai and Tongde county)	10.67	6.54	13,476	2043.83	316.4	291.52
Huangnan (only Zeku and Henan county)	8.40	8.01	16,165	1355.49	201.4	409.07
Haixi (Only Tanggula town)	0.10	0.10	314	1730.23	10.63	14.73

Global warming and intensified evaporation have promoted reverse succession. Wang et al. (2000, 2001) attribute the large-scale degradation of alpine meadow and steppe vegetation to the warming of the regional climate over the last 40 years, largely associated with the degradation of previously frozen land. Similarly, Chen et al. (1998) suggest that global warming has induced desertification that has degraded grassland areas. This warming trend has impacted upon plant growth, yield and community structure in alpine meadow ecosystems (Li et al. 2004). Extreme weather and climate disasters in the 1990s intensified grassland degradation.

Figure 6.5a shows significant increases in average annual temperature in the Sanjiangyuan region from 1961 to 2004 (Li et al. 2006; Wang 2007). This increase was especially marked in the 1980s and 1990s. Mean annual temperature was −2.36°C in the 1960s, −2.19°C in the 1970s, −2.04°C in the 1980s, and −1.78°C in the 1990s. Mean annual temperature in 2003 was the highest on record, 1.4°C higher than the average for the previous 30 years. As temperature adjustments are most marked in autumn and winter, the annual variation of temperature has decreased year by year (Li et al. 2006; Fu et al. 2007b; Chen et al. 2007). This has induced glacier retreat, ascending snow lines, drying up of wetlands, and degradation of alpine permafrost.

Mean annual precipitation in the Sanjiangyuan region from 1961 to 2004 is shown in Fig. 6.5b (Li et al. 2006; Wang 2007). While the 1980s were characterized by slightly above average precipitation, this trend was reversed in the 1990s. Seasonal changes have varied across the region (Chen et al. 2007). Major snowstorms occurred in the Sanjiangyuan region during winters of the late 1990s and in the spring early in the twenty first century.

Mean annual evaporation in the Sanjiangyuan region increased slightly from 1961 to 2004, with an average increase of 0.13 mm a^{-1} (Fig. 6.5c). Mean annual evaporation was notably lower through the 1980s, but increased in the 1990s (Wang 2007; Chen et al. 2007). To some degree, climate warming has reduced the inhibiting effect of low temperatures upon plant growth in alpine meadows, but evaporation of ground surface and evapotranspiration from vegetation has increased faster than precipitation, such that water availability is the primary limiting factor for plant growth (Li et al. 2000). Intriguing debates are emerging about the relative impacts of climate change and human impacts on the vegetation structure and function atop the Qinghai-Tibetan Plateau (cf., Klein et al. 2007; Miehe et al. 2008, 2011).

6.3.2 Grassland Degradation and Desertification

The influence of global warming, intensified human activity, overgrazing, and frequent natural disasters have degraded almost 90% of natural grassland areas in Qinghai province in recent decades (Department of Animal Husbandry and Veterinary 1996; Fu et al. 2007a; cf., Harris 2010, Li et al. 2011). This has markedly decreased the primary productivity associated with animal husbandry in

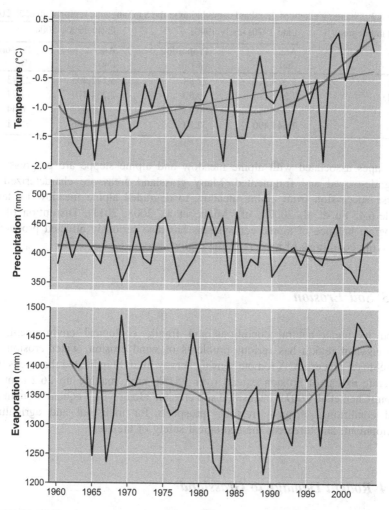

Fig. 6.5 Variability in mean annual temperature (**a**, *upper graph*), precipitation (**b**, *middle graph*) and evaporation (**c**, *lower graph*) in the Sanjiangyuan region (1961–2004; data from Li et al. 2006; Fu et al. 2007a, b; Chen et al. 2007; Wang 2007)

these areas, threatening people's livelihood. Remote sensing analysis indicates that the area of degraded grassland increased from 764×10^4 ha in the late 1970s— early 1990s (32.8% of the total region) to 841×10^4 ha in the period 1990s—2004 (36.11% of the total region) (Table 6.3, Liu 2008). Moderately degraded grass- lands now extend across around 6.9 million ha, which is 36% of the total utilizable grassland (Chen et al. 2007). Compared to 1950, the yield per unit area of grass- land has decreased by 33%, the percentage of elite forage species has decreased by 25% and vegetation cover has decreased by 20%, while the percentage of toxic plants has increased by 75%, the height of dominant plants has decreased by 40%, and the height of grasses has decreased by 20% (Fu et al. 2007a). Ecological

Table 6.3 Changing extent of grassland degradation in the Sanjiangyuan region (Liu et al. 2008)

Grassland degree	Late 1970s–early 1990s		Early 1990s–2004	
	Area (ha)	Proportion (%)	Area (ha)	Proportion (%)
Slight	5,328,317	22.88	5,572,405	23.93
Middle	2,212,216	9.50	2,734,779	11.74
Severe	103,957	0.45	103,082	0.44
Total	7,644,490	32.83	8,410,266	36.11

landscapes associated with alpine meadow and alpine steppe are progressively being destroyed and fragmented. Many grassland areas are characterized by reverse succession from alpine meadow to degraded alpine meadow to desert (Table 6.4; Fu et al. 2007b; cf., Miehe et al. 2008, 2011). Disruption of the ecological balance of alpine eco-systems has also seriously affected production and living conditions of herders in the region.

6.3.3 Soil Erosion

Given its harsh natural conditions and fragile ecological environment, the Sanjiangyuan region has serious problems of wind erosion, water erosion and freeze–thaw processes. A remote sensing survey in 2000 indicated that areas with moderate extent of erosion extend over 95,000 km², making up 26.2% of the region (Chen et al. 2007). Intensified soil erosion, along with frequent drought and flood conditions, seriously restricts prospects for industrial and agricultural development, and threatens the ecological safety of the area.

6.3.4 Rodent Damage to Grassland

Although the plateau pika (*Ochotona curzoniae*), plateau zokor (*Myospalax baileyi*) and plateau vole (*Pitymys irene*) are native mammals, they are referred to as '*rodents*' of the Qinghai–Tibetan Plateau. Grasslands are severely impacted by their burrowing and gnawing behavior (Zhou et al. 2003a, b, 2005). Their activities accelerate erosion and degradation rates by loosening the Kobresia sod and killing its roots (Limbach et al. 2000; Zhou et al. 2005). The area affected by these burrowing animals in the Sanjiangyuan region is 6.4 million ha (Chen et al. 2007). This makes up 17% of the total area of the region, and 33% of the total utilizable grassland. As many as 374 pikas/ha have been recorded in some areas (Ma et al. 2000), with up to 1335 burrows/ha (Zhou et al. 2005). Natural enemies for burrowing animals have been greatly reduced since the 1980s because of illegal hunting (especially eagles), impacting greatly upon the food chain. The 'degradational' role of these burrowing mammals remains contentious. Indeed, some argue that some of

Table 6.4 Increase in desertification area in the Sanjiangyuan region (1949–1998) (Fu et al. 2007a)

Year	Yellow river (km²)			Yangtze river (km²)			Data source
	Area	Increased area	Annual average increased area	Area	Increased area	Annual average increased area	
1949	1961.00			7430.60			
1959	2208.90	247.90	24.79	8369.94	939.34	93.93	Aerial photographs supplied by the Sand control team of Qinghai
1977	2923.00	714.10	39.67	11075.80	2705.86	150.33	Aerial photographs supplied by the Forest investigative team of Qinghai
1986	3540.90	617.90	68.66	13417.14	2341.34	260.15	Fu et al. 2007a
1994	4636.10	1095.20	136.90	17567.06	4149.92	518.74	Satellite data
1998	5357.60	721.50	180.38	20300.96	2733.90	683.48	Satellite data

Table 6.5 Average annual river runoff in the Sanjiangyuan region the late twentieth century ($\times 10^8$ m^3; Han 2004)

Region	1956–2000	1991–2000
Yellow source	204.9	174.8
Yangtze source	122.1	112.0
Mekong source	43.5	42.9
Sanjiangyuan region	370.2	329.7

these species, especially plateau pika, are critical ecosystem engineers in this region (see Smith and Foggin 1999).

6.3.5 Declining Water Supply From Headwater Areas

Water production in headwater areas of Qinghai province has declined in recent years. Upper reaches of the Yellow River stopped flowing for the first time in recorded history in 1997 (Han 2004). Since 1990, water levels of Zhaling Lake and Eling Lake have declined by 2 m, and under extreme conditions flow between the lakes ceased. Around 1070 lakes in the Sanjiangyuan region have dried up in the last 20 years, of which 1040 were in the Yellow River source zone (Chen et al. 2007). From 1956 to 2000 average annual river runoff in Sanjiangyuan was 370.20×10^8 m^3, but it decreased to 329.70×10^8 m^3 in 1991–2000, a reduction of 10.90% (Table 6.5). The prolonged decline in discharge in the upper reaches not only constrains the sustainable development of herders living in the region itself, it also influences the livelihood of people downstream. The retreat of glaciers and shrinkage of lakes has caused a decrease in water level and degradation of wetlands in the region. Reduced runoff has brought about a water supply crisis in some settlements and towns. Herders have been forced into other areas in efforts to enhance their living standards, increasing grazing pressure and further degrading grassland areas (Chen et al. 2007).

6.3.6 Biodiversity Losses

Habitat fragmentation and isolation have brought about biodiversity losses in the region. Threatened species make up 18% of the total number of species, much higher than the world average of 13% (Chen et al. 2007). As many living forms on the plateau have special germplasm, the prospective disappearance of these species would reduce the gene pool that has adapted to the frigid alpine conditions. Habitat fragmentation and illegal hunting have reduced numbers of protected wildlife such as Tibetan antelope from a population of 10×10^4 in 1985 to 3×10^4 in 2002 (subsequent increases reflect tightening of regulations and anti-poaching efforts). Alpine musk deer (*Moschus sifanicus*) are almost extinct (Chen et al. 2007); these animals are highly sought for musk – a traditional medicine. There have also been declines in national first class protection animals such as white-lipped deer (*Cervus albirostris*), red deer (*Cervus elaphus*) and snow leopard.

Table 6.6 Population, livestock and pasture changes in Qinghai province (1949–2003) (Fu et al. 2007b)

Year	Population of pastoral area (10^4 persons)	Livestock number (10^4 Cattle/sheep)	Livestock number per capita (Cattle/sheep)	Grassland area per capita (ha)	Grassland area per livestock (ha)
1949	21.96	748.73	34.10	143.87	4.20
2003	70.00	2217.65	31.68	45.13	1.40

6.3.7 Human Induced Factors that Promote Ecological Degradation in the Sanjiangyuan Region

Socio-economic development in the Sanjiangyuan region relies largely upon grazing livestock, supplemented by activities such as illegal hunting, uncontrolled gathering and digging of herbal medicine, and mining. These harmful practices and ineffective management of resources have brought about numerous environmental problems. Population growth has increased the demands for animal husbandry in Qinghai, indirectly affecting grassland areas. Alongside climate warming and economic activities, severe ecological degradation in the region has resulted, in part, because of grassland property-rights, law regulation and cultural transformations (Ma 2007). Grassland property-rights in the Sanjiangyuan region have been transformed from an early tribe and temple (or monastery) possession system through a mutual cooperation system from 1949 to 1958, a community possession system from 1958 to 1978, and a family contract management responsibility system from 1978.

Bai et al. (2002) consider overgrazing to be the key reason for grassland degradation in Maduo County. Since the 1950s the human population of pastoral areas of Qinghai province has increased by 3 times, and the number of livestock has more than doubled (Table 6.6; Fu et al. 2007a). In response, grassland area per sheep unit has decreased from 4.20 ha in 1949 to 1.40 ha in 2000. This doubling of the stocking rate has forced herders to graze their livestock at even higher elevations, expanding the impact and damage of these activities (Wang and Cheng 2001; Zhou et al. 2003b, 2005). As shown on Fig. 6.6, stocking rates increased notably in the 1990s, reaching a peak in 1992 (Fu et al. 2007a). The theoretical capacity of stocking on natural grassland in Qinghai was 36,254,500 sheep units in the early 1980s (Fu et al. 2007a). Grassland degradation has reduced the current carrying capacity of grassland to 75% of that experienced in the 1980s (around 27,200,000 sheep units; Agricultural Resources Section Office of Qinghai Province 1999). Based on these numbers, the region was overstocked by about 6,330,000 sheep units from 1980 to 2003. This effect was most pronounced from 1990 to 1996, when the average rate of overload was 33.6% (Fig. 6.6). This is consistent with the accelerated rate of grassland degradation at this time (Fu et al. 2007a). Grassland overgrazing is most pronounced in winter and spring. It has brought about sparse, low and degraded cover and enhanced growth of poisonous plants, reducing the capacity for self-restoration. These consequences, in turn, provide good conditions for burrowing animals. Areas of alpine meadow have become extremely degraded as the vegetation community

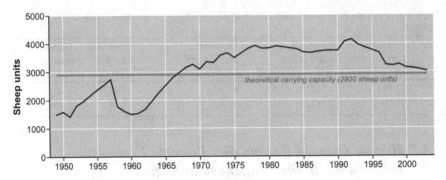

Fig. 6.6 Trend in sheep units ($\times$ 10^4) in the Sanjiangyuan region (1949–2004), showing the excess stocking rate since the 1970s (Fu et al. 2007b)

structure has been altered and soil fertility has been depleted. A vicious circle has been generated, characterized by overgrazing—vegetation degradation—harm by rodents and pests—grassland degradation—intensified mismatches between animal husbandry and management of grasslands.

Qinghai is a province rich in wildlife resources, with numerous unique wildlife of high economic value. Since the 1980s, illegal hunting has become a more and more serious problem (Wang et al. 2000; Zhou et al. 2005). More than 100,000 musk deer were killed in the 1980s, reducing their population by 90%. Snow leopards are increasingly rare. Illegal hunting resulted in the loss of 32,000 Tibetan antelope at the end of 1990s (Chen et al. 2007). The on-going illegal hunting of wild yak and wild ass (*Equus kiang*) has resulted in the rapid reduction of this rare and valuable wildlife.

Economic interests have encouraged more than 200,000 outsiders to move into the Sanjiangyuan region seasonally for gold mining or gathering of herbal medicines. This has resulted in serious damage to vegetation. In Autumn–Spring 2000, around 200 ha/day of shrub grassland was damaged and 7–8 m^2/day of grassland sod was destroyed adjacent to herders dwellings (Chen et al. 2002). Local herders use livestock manure and roots and stems of plants as a fuel, damaging grassland. In addition, gold mining disturbed 1.07 million ha of grassland in the 1980s, of which 33,000 ha was thoroughly destroyed (Zhou et al. 2005). Similar rangeland management issues are being addressed in the North-West of China (Squires et al. 2010).

6.4 Programme to Protect and Restore Socio-Economic and Environmental Values in the Sanjiangyuan National Nature Reserve

Efforts to protect biodiversity while enhancing the regional economy and prospects for development prompted the establishment of the Sanjiangyuan National Nature Reserve. The reserve comprises an area of 152,300 km^2, making up 21% of the

Fig. 6.7 Distribution of core (*dark tone*), buffer (*middle tone*) and experimental (*light tone*) areas in the Sanjiangyuan National Nature Reserve

total land area of Qinghai province and 42% of the Sanjiangyuan region (Chen et al. 2007). Protection and restoration of high altitude plateau ecosystems is a critical step in efforts to provide abundant, high quality water resources for the Yellow, Yangtze and Mekong Rivers. Large scale planning and design are required to reverse degradation trends in the region. This entails both protection on the one hand, and restoration of forest, grassland and wetland ecosystems on the other. The main protection targets are:

- Alpine wetland ecosystems, including glaciers and snow-covered mountains, swamps, and lakes;
- Protected wildlife and other species of national and provincial importance, including Tibetan antelope, wild yak, snow leopard, blue sheep, Tibetan gazelle, caterpillar fungus and orchids (*Orchidaceae* spp.);
- Sparse alpine forest ecosystems, such as Balffour spruce (*Picea likiangensis var.*) forest and Qilian savin (*Sabina przewalskii*) forest.

The functional area of the Sangjiangyuan Natural Reserve is divided into core, buffer and experimental areas (Fig. 6.7). Core areas are strictly protected areas; buffer areas are important protected areas; experimental areas are normal protected areas in which consideration is given to both protection and utilization. The 18 core areas take up 31,218 km^2, equivalent to 20.5% of the total land area of the reserve (Chen et al. 2007). The present human population within core areas is 43,566 people. Criteria used to design the core areas include:

- To protect typical natural ecosystems in the area, fostering growth and repro-
  duction of targeted wildlife, plants and organisms and their habitats.
- To separate areas of environmental protection and restoration from human
  activities.

Landscape planning and GIS analysis used population viability analyses to
designate the network of core zones and associated corridors (Tang 2003;
Table 6.7). Three core areas, Suojia-Quma River, Jiangxi and Baizha, which take
up 37% of the total core area, have been designated to protect wildlife. Seven core
areas, Tongtianheyan, Dongzhong, Angsai, Zhongtie-Jungong, Duoke River,
Maixiu and Make River, taking up 15% of the total core area, protect typical forest
and shrubland. Finally, eight core areas, Animaqin, Xingxingha, Nianbaoyuze,
Dangqu, Geladandong, Yueguzonglie, Erlin-Zalin Lake and Gouzongmucha,
taking up 48% of the total core area, are used to protect wetland ecosystems (Chen
et al. 2007). Core areas in the west typically focus on wildlife, core areas in the
east focus on forest and shrubbery, and core areas in headwater areas and around
lakes focus on wetland ecosystems. Protection measures include closing the area
for strict protection, forbidding hunting, suspending grazing, stopping deforestation, and forbidding resource development activities.

Buffer areas surround core areas, or they connect core areas to assist in
protecting targets (i.e. they address concerns for fragmentation; see Fig. 6.7).
The principle tasks of the buffer areas are to control the impact of threatening
factors/processes, and to restore and harness slightly degraded ecosystems. Buffer
areas take up 39,242 km^2 (25.8% of the total land area of the reserve). The present
human population in buffer areas is 54,254 people (Chen et al. 2007). Criteria used
to design buffer areas include:

- To buffer main protection targets from influences outside the natural reserve.
- To link core areas to assist in the protection of wildlife.
- Separation from towns, factories and mining sites.

Measures taken in the buffer areas include reducing grazing livestock numbers
to sustainable forage, controlling grazing intensity through rotational grazing, and
closing some areas for restoration of forest and grassland vegetation.

Experimental areas outside core and buffer areas take up 81,882 km^2 (53.7% of
the total land area of the reserve; Fig. 6.7). The present human population in
experimental areas is 125,270 people (Chen et al. 2007). Criteria used to design
experimental areas include:

- To aid the development and improvement of socio-economic conditions, production and living standards of herders, while promoting societal progress by
  adjusting industrial structures to optimize the disposition of resources and
  promote regional opportunities (e.g. eco-tourism).
- To assist the restoration and rebuilding of degraded ecosystems.
- To enhance the management of fragmented protection targets by providing a
  natural defence for core and buffer areas.

Table 6.7 Protection aims in the 18 core areas of Sanjiangyuan National Nature Reserve (Tang 2003)

Core name	Protection targets	Location	Area (km²)
Animaqin	Snow mountains and glaciers	North–West of Maqin county	507
Xingxinghai	Lakes and wetlands	In Madou county	984
Nianbaoyuze	Snow mountains, glaciers and lakes below	In Jiuzhi county	262
Dangqu	Wetlands and swamps	West of Zadou county	5,843
Geladandong	Glaciers and vegetation around	Tanggulashan town in Geermu city	1,952
Yueguzonglie	River, lakes and swamps	In Madou town of Qumalai county	963
Erlin–Zalin Lake	Lakes and swamps around	In Madou county	1,818
Gouzongmucha	Snow mountains river lakes and swamps	Between Zadou and Zhidou county	2,883
Soujia-Quma River	Tibetan antelopes wild yaks, monkeys etc. and swamps	Between Qumalai and Zhidou county	10,684
Jiangxi	Macaques etc. wild animals and their habitat	In Nangqian and Yushu county	337
Baizha	Golden leopards, snow leopards and clouded Leopards	In Nangqian county	419
Tongtianheyan	Cypress and shrub forests	In Yushu and Chendou county	1,355
Dongzhong	Spruce forests	In Yushu county	493
Angsai	Forests and shrubs	In Zadou county	356
Zhongtie-Jungong	Spruce forests	Between Maqin, Tongde and Xinghai county	1,341
Duoke River	Conifers	In Banma county	110
Maixiu	Forests	In Zeku county	544
Make River	Defoliate broadleaf, conifers and shrubs	In Banma county	367
Total			31,218

Projects planned for experimental areas include:

- In areas where human population is greater than the carrying capacity, herders will be resettled in aggregated locations. Prospectively, this ecological immigration policy will minimize pressure on the natural grassland, enhancing restoration efforts.
- To determine sustainable grazing livestock numbers given the quantity of available forage in the area, suspending grazing in some areas in efforts to restore grassland vegetation.
- To protect and rehabilitate forest and grassland vegetation, wetlands and wildlife.
- To construct water supply facilities and develop new energy for household needs.
- To set up research and monitoring bases.

Planning of the Sanjiangyuan National Nature Reserve considers ecology-based protection and construction in relation to herders production and lifestyle values while accelerating the development of the local economy. Proactive, preventative measures taken to improve ecological conditions include suspension of grazing to improve grassland vegetation, reverting crop land back to forest or grassland cover, and fencing off areas for restoration of grassland vegetation. This is accompanied by resettlement of ecological immigrants into small towns, rodent-disaster programmes and small-scale water and soil erosion measures in efforts to enhance livestock production and control the spread of desert and areas of degraded grassland (see Foggin 2008). These initiatives will assist efforts to promote water conservancy (Zhou and Yie 2007). However, experience shows that such programmes are not always effective (see Agrawal and Redford 2009; Dowie 2009; Blue 2010; Foggin 2011).

Measures taken to improve wetland conditions will include:

- enclosing wetland areas to promote self-regeneration of natural vegetation
- re-sowing wetlands that have been subjected to severe degradation
- reducing livestock numbers or stopping grazing adjacent to wetlands
- application of biological and ecological techniques to control rodents in areas where they have brought about severe damage.

Bans on fishing and wildlife management programmes protect animal resources in the region. Marked improvements in energy sources will improve herders' income and reduce dependencies upon remaining areas of forest and natural grassland. New technologies will be developed and trialed, such as rainfall generation (especially in the Yellow River source area) and engineering practices that will reduce the threat of desertification. International experience guards against undue expectations from such 'technofixes' (e.g. Higgs 2003). Hazard reduction management will include attempts to fire proof areas of forest and grassland. Ecological compensation mechanisms will be put into effect for those people affected by these environmental protection and restoration programmes (Li 2007).

Prospects for ecologically-linked industrial and societal developments in the Sanjiangyuan region are contingent upon protection of ecological values in the first instance. The main conflict between human and ecological values in the region relates to the protection/repair of ecosystem attributes in the face of increasing demands for land and water use brought about by population growth (Wang and Wang 2004). Of the 651,000 people in the region, 90% are Tibetan (Sheng 2006). Survival of herding traditions requires utilization and income generation from local resources in a manner that minimizes harm to the environment through overgrazing, soil erosion, desertification, and reduced water availability. Migration policies have been applied in light of these population pressures. These measures have been supplemented by assistance to develop more intensive industries, along with training programmes to improve labour skills (Sheng 2006). Steps taken include: modernization of the animal husbandry industry (including enhanced livestock processing; Zhang et al. 2005), development of regional grassland industries, production of new Chinese and Tibetan medicines, and promotion of ecotourism opportunities (Fang and Liu 2006; Wang et al. 2007). Animal husbandry production is a main component of economic development in the Sanjiangyuan region. As traditional practices applied in the farming of yak and Tibetan-sheep use no fertilizer, pesticide or other chemically synthesized materials, the Sanjiangyuan region is an ideal area to develop organic agriculture (Guo 2005; Shi 2006). These prospects promote a win–win 'green economy' for the plateau, protecting the environment while improving farmer's income. To assist these developments, adjustments to the service sector are underway, including infrastructure and governance arrangements, transportation, education and monitoring programmes (Chen et al. 2007).

The total investment for the Program on Ecology-based Protection and Construction of the Sanjiangyuan National Nature Reserve in Qinghai is estimated at 7.5 billion Yuan (RMB) (Chen et al. 2007). This includes:

- 4.92 billion Yuan for ecology-based protection and construction.
- 2.22 billion Yuan for construction of infrastructure to enhance living standards for herders.
- 0.36 billion Yuan for measures to support ecological protection.

6.5 Discussion

Efforts to address the 'big' questions, rather than becoming overwhelmed by the tyranny of small-scale thinking and approaches, have prompted the development of numerous large-scale conservation initiatives in differing parts of the world. These landscape-scale applications recognize that extensive reserves and/or protected areas are required to maintain and/or enhance the resilience of ecosystems. Reserve systems across the world comprise a biased sample of biodiversity as they are typically located in remote places and in areas that are relatively unsuitable for commercial activities (Pressey 1994). Having said this, mountain regions are not only treasures of high biodiversity that are rich in endemic species, they also provide a fundamental source of high quality water, other products (wood, minerals, game, food, traditional medicines, etc.), and recreational opportunities (Hamilton and Macmillan 2004). Unless these areas are protected as reserves, human exploitation of resources is likely to occur, and degradation will ensue. Reserves are required to support long-term survival of species (e.g. Soulé 1987). In most cases, conservation goals will not be met unless the ecological integrity of conservation targets is restored (Hargrove et al. 2002). Ecological restoration is expensive, costing an estimated 2–5 times more than conserving intact and viable examples of natural communities (The Nature Conservancy 2000). Conservation planning, as such, is a fundamental and cost-effective component of biodiversity management and sustainable development.

Biological diversity is inextricably linked to the variety of landscapes in any ecoregion. Healthy ecosystems are self-sustaining and resilient–evolving systems that adapt and change over time. In fragmented landscapes populations can be prevented from reaching migration and dispersal destinations such that they are forced to live in habitats that are not large enough for their survival as they are unable to achieve genetic exchange. The steep environmental gradients and proximity of different altitudinal zones in mountainous regions present significant opportunities for biotic adaptation to environmental changes. Hence, large areas must be conserved to protect the adaptive capacity of these systems, giving species the opportunity to migrate to new habitats. The Sanjiangyuan National Nature Reserve is among the largest nature reserves in the world, with the highest and most extensive wetland protected area. The total area of the reserve is marginally greater than the land mass of England and Wales.

Water resources planning strategies and landscape ecology programmes in the Sanjiangyuan National Nature Reserve link lake and river ecosystems to grassland, forest and wetland management programmes at the landscape scale. The 18 core areas of the reserve have been framed in relation to water management strategies in wetlands, lakes and forests (Tang 2003; Chen et al. 2007; see Table 6.7). In this era of environmental repair, it is recognized explicitly that conservation programmes alone are not enough, and major efforts have been put in place to facilitate rehabilitation, promoting self-recovery mechanisms wherever possible (Brierley and Fryirs 2008). An ecosystem approach to environmental management has been adopted, wherein these applications strive to 'work with nature', emphasizing big-picture relationships at the 'whole of system' scale.

The approach to conservation planning adopted in the Sanjiangyuan National Nature Reserve mimics developments elsewhere in the world, emphasizing concerns for targeted species within process-based conservation and restoration plans. An umbrella species approach to conservation focuses upon keystone species, thereby facilitating the conservation of many other species with coincident, or a smaller range of, habitat needs. The approach to ecological protection and restoration in the Sanjiangyuan National Nature Reserve closely parallels the aspirations of the Pan-European Ecological Network (STRA-REP 2006) in that it aims to:

• Conserve a full range of ecosystems, habitats, species and landscapes in efforts to ensure the existence of healthy, intact and connected wildlife habitat.

- Ensure landscapes are large enough to conserve viable populations of keystone species.
- Reduce the impacts of fragmentation, providing sufficient opportunities for the dispersal and migration of species. This enables genetic exchange between different local populations and allows local populations to move away from degraded habitats.
- Restore damaged parts of key environmental systems, while securing the integrity of vital environmental processes.
- Buffer key environmental systems from potential threats.

Linked networks of core areas, corridors and buffer zones enhance resilience and mitigate against biodiversity loss and impairment of ecosystem services, promoting flexibility and adaptivity in light of environmental changes. Core areas with a variety of linkage zones are required to fulfill the vision of contiguous wildlife habitat (Bouwma et al. 2002). These areas act as reservoirs for biodiversity where evolution and reproductive processes can take place. A physical network of core areas, linked by corridors and supported by buffer zones, enables the dispersal and migration of species. Linkage zones are more than simple corridors: they are habitats with sufficient food and resources to support minimal wildlife populations and to accommodate the continual residence and movement of species between core areas (Bouwma et al. 2002). Corridors facilitate ecological connectivity and coherence, thereby enhancing the self-regulating capacity of ecosystems as species are able to disperse, migrate, forage and reproduce. Buffer zones provide protection from adverse external damages and disturbance, increasing the flexibility and adaptive capacity of core areas to respond to environmental changes. These areas may perform a corridor function or in themselves harbour valuable biodiversity, such as species populations that are dependent on traditional forms of agriculture (Bouwma et al. 2002). Sound management of human activities in these areas decreases the potential impacts and probability of isolation. Buffer areas that allow for approved manipulative research and low impact uses by residents surround well-protected core areas in the Sanjiangyuan National Nature Reserve.

Effective conservation programmes do not view ecological reserves as isolated landscape fragments. Rather, protected areas are linked to the wider region, providing a basis for the sustainable use of landscapes and natural resources. Precautionary measures provide sufficient ecological and landscape quality over as wide a range of territory as possible. This enhances the adaptive capacity of the system, enabling appropriate adjustments to be made in response to social and environmental changes. Many species within protected areas depend upon resources outside them. The transitional zone (experimental area) beyond the core/buffer areas in Sanjiangyuan National Nature Reserve has been established to promote collaboration and engagement with local residents in efforts to meet the conservation goals of the reserve (c.f., Smardon and Faust 2006).

Monitoring networks have been set up in the Sanjiangyuan region to support decision-making for further utilization and conservation and to adjust policies on the reserve are required. For example, GIS, remote sensing and modeling applications are being used to assess the rate and extent of damage by burrowing mammals, grassland degradation, desertification and the pattern/rate of erosion. Various experimental programmes have been established to trial new approaches to improve the management of wetlands, desertification, fish stocks in lakes, regeneration of grasslands, etc. Recent research has highlighted the success of these measures. There are increased numbers of wild animals such as blue sheep, wild ass, wolves (*Canis lupus*), golden eagle (*Aquila chrysaetos*), Tibetan antelopes and snowcocks (*Tetraogallua tibetanus*) (Ren 2008). Lakes that had dried up have water once more (Zhu and Dai 2008). Vegetation cover has reestablished in many areas of bare land (Zhu and Dai 2008). Critically, these elements work together to enhance ecosystem health. For example, recovered wild animals help to re-establish the food chain structure of the plateau ecosystem, as eagles and weasels limit undue degradational influences of burrowing mammals upon grassland (recognizing, in turn, that plateau pika are themselves key ecosystem engineers (see Smith and Foggin 1999)). Results from environmental monitoring programmes will be utilized to assess trends in ecosystem performance, identifying threatening processes that may compromise the integrity of ecosystems (cf. Margules and Pressey 2000).

Visions for an ecologically sustainable future integrate proactive biodiversity management programmes with coherent strategies that promote regional development and natural resource management. The quest for sustainability frames environmental condition in relation to socio-economic and cultural considerations—both now and into the future. Ecological degradation and biodiversity losses in the Sanjiangyuan region will continue unless human developments are managed appropriately (Foggin et al. 2006). Community engagement and education are required to unite conservation goals with sustainable development initiatives. Available resources in the Sanjiangyuan region are unable to meet the needs of the growing population, creating an imbalance between human development and environmental resources. Measures adopted to address this problem include migration policies (Sheng 2006; Chen et al. 2007), increasing local economies by developing ecotourism opportunities (Fang and Liu 2006; Wang et al. 2007), enhanced production of organic foods (Guo 2005; Shi 2006) and increasing high technology applications (Chen et al. 2007; Li 2007a, Zhang et al. 2005). Importantly, benefits from environmental conservation and rehabilitation measures extend well beyond the Sanjiangyuan region. Coherent approaches to management of water resources in headwater areas assist prospects for stable and sustained economic development downstream. Programmes that facilitate environmental protection and restoration in the Sanjiangyuan region promote ecological security for half of China.

Acknowledgments This research was supported by the National Natural Sciences Foundation of China (30760160), MOE (2010-1595) and MOST 2011DFA20820. The authors thank Chen Gang, Jiang Lu-Jia, Qiao You-ming, Li Ji-lan, Pei Hai-kun, Lu Guang-xin, Wang Chun-qing, Fu Yang and Zhang Jing for their help in collecting information for the manuscript, and Igor Drecki for drafting the figures. Helpful guidance by the editor of this special issue, David Higgitt, and two anonymous reviewers is gratefully acknowledged. Critical appraisal and constructive guidance by Marc Foggin is particularly appreciated.

References

Agrawal A, Redford K (2009) Conservation and displacement: an overview. Conservat Soc 7(1):1–10

Agricultural Resources Section Office of Qinghai Province (1999) Agricultural resources data, Qinghai, China. pp 100–164 (in Chinese)

Bai W, Yang Y, Xie G, Shen Z (2002) Analysis of formation causes of grassland degradation in Maduo county in the source region of yellow river. Chin J Appl Ecol 13(7):823–826 (in Chinese with English abstract)

Blue L (2010) International conservation trends: considerations for Sanjiangyuan National Nature Reserve. In: Chen G, Li X, Gao J, Brierley G (eds) Wetland types and evolution and rehabilitation in the Sanjiangyuan region. Qinghai People's Publishing House, Xining, pp 111–128

Bouwma IM, Jongman RHG, Butovsky RO (eds) (2002) The indicative map of Pan-European ecological network–technical background document. ECNC technical report series, ECNC, Tilburg

Brierley G, Fryirs K (eds) (2008) River futures: an integrative scientific approach to river repair. Island Press, Washington

Cannon KA (2006) Water as a source of conflict and instability in China. Strateg Anal 30:310–328

Chape S, Harrison J, Spalding M, Lysenko I (2005) Measuring the extent and effectiveness of protected areas as an indicator for meeting global biodiversity targets. Philos Trans Royal Soc B 360:443–455

Chen G, Chen X-Q, Gou X-J (eds) (2007) Ecological environment in Sanjiangyuan natural reserve. Qinghai People's Press, Xining (in Chinese)

Chen Q, Liang T, Wei Y (1998) Causes of grassland degradation in Dari county of Qinghai Province. Acta Pratacultural Sci 7(4):44–48 (in Chinese with English abstract)

Chen X, Gou X (eds) (2002) Ecological environment in Sanjiangyuan natural reserve. Qinghai People's Press, Xining (in Chinese)

Department of Animal Husbandry, Veterinary, Institute of Grassland of the Chinese Academy of Agricultural Sciences, Commission for Integrated Survey of Natural Resources of the Chinese Academy of Sciences. (1996) Data on grassland resources of China. China Agriculture Science and Technology Press, Beijing (in Chinese)

Dowie M (2009) Conservation refugees the hundred-year conflict between global conservation and native peoples. MIT Press, Cambridge, Massachusetts

Du M, Kawashima S, Yonemura S, Zhang X, Chen S (2004) Mutual influence between human activities and climate change in the Tibetan Plateau during recent years. Glob Planet Chang 41:241–249

Economy E (1999) China's Go West Campaign: ecological construction or ecological exploitation. China Environ Ser 5:1–10

Fang C, Liu H (2006) The eco-industry development strategy and demonstrating-district construction in Sanjiangyuan. J Mt Sci 24(6):744–760

Foggin JM (2008) Depopulating the Tibetan grasslands: national policies and perspectives for the future of Tibetan herders in Qinghai Province, China. Mt Res Dev 28(1):26–31

Foggin JM (2011) Rethinking "ecological migration" and the value of cultural continuity: a response to Wang, Song and Hu. Ambio 40(1). doi:10.1007/s13280-010-0105-5

Foggin PM, Torrance ME, Dorje D, Xuri W, Foggin JM, Torrance J (2006) Assessment of the health status and risk factors of Kham Tibetan pastoralists in the alpine grasslands of the Tibetan plateau. Soc Sci Med 63(9):2512–2532

Fu Y, Li F, Zhang G, Yang Q, Zeng X (2007a) Natural grasslands degradation and environmental driving factors in Qinghai Province. J Glaciol Geocryol 29(4):525–535 (in Chinese with English abstract)

Fu Y, Zhang G, Li L (2007b) The change features and its tendency of ecological environment on Qinghai-Tibetan Plateau. J Qinghai Meteorol 2:4–10 (in Chinese with English abstract)

Goodman DSG (ed) (2004) The Campaign to 'Open Up the West': national, provincial-level and local perspectives. Cambridge University Press, Cambridge

Guo Y (2005) Countermeasures and proposals on organic animal husbandry development in Qinghai Province. Qinghai Prataculture 2:32–34 (in Chinese with English abstract)

Hamilton L, Macmillan L (eds) (2004) Guidelines for planning and managing mountain protection areas. IUCN World Commission on Protected Areas, Gland

Han Y (ed) (2004) The evaluation of hydro-resource in Qinghai. Qinghai People's Press, Xining (in Chinese)

Hargrove B, Tear T, Landon L (2002) Geography of hope update. When and where to consider restoration in ecoregional planning. www.earthscape.org/p1/ES14454/. Accessed Nov 2007

Harris RB (2010) Rangeland degradation on the Qinghai-Tibetan plateau: a review of the evidence of its magnitude and causes. J Arid Environ 74:1–12

Higgs E (2003) Nature by design. People, natural process, and ecological restoration. MIT Press, Cambridge, Massachusetts

Klein JA, Harte J, Zhao X-Q (2007) Experimental warming, not grazing, decreases rangeland quality on the Tibetan Plateau. Ecol Appl 17(2):541–557

Li X-L, Gao J, Brierley G, Qiao Y-M, Zhang J, Yang Y-W (2011) Rangeland degradation on the Qinghai-Tibet plateau: implications for rehabilitation. Land Degradation & Development 22:1–9

Li J (2007a) A research on ecological protection and construction of the Sanjiangyuan region in Qinghai. Shanxi For Sci Technol 3:58–59 (in Chinese with English abstract)

Li K (2007b) A protection proposal of water conservancy engineering on fish production and hydrophily living things. China Fish 12:72–73 (in Chinese with English abstract)

Li Y, Wang Q, Zhao X (2000) The influence of climatic warming on the climatic potential productivity of alpine meadow. Acta Agrestia Sinica 8(1):23–29 (in Chinese with English abstract)

Li Y, Zhao X, Cao G (2004) Analyses on climates and vegetation productivity background at Haibei alpine meadow ecosystem research station. Plateau Meteorol 23(4):558–567 (in Chinese with English abstract)

Li Y, Li F, Guo A, Zhu X (2006) Study on the climate change trend and its catastrophe over Sanjiangyuan region in recent 43 years. J Nat Resour 21(1):79–85 (in Chinese with English abstract)

Limbach W, Davis J, Bao T, Shi D, Wang C (2000) The introduction of sustainable development practices of the Qinghai livestock development project. In: Zheng D, Zhu L (eds) Formation and evolution, environment changes and sustainable development on the Tibetan Plateau. Academy Press, Beijing, pp 509–522

Liu J, Xu X, Shao Q (2008) The spatial and temporal characteristics of grassland degradation in the three-river headwaters region in Qinghai Province. Acta Geogr Sinica 63(4):364–376

Ma H (2007) Explanation about the causes of ecological degradation in the Sanjiangyuan District under the perspectives of new institutional economics. Tibet Stud 3:88–96 (in Chinese with English abstract)

Ma Y, Lang B, Shi D (2000) Establishing pratacultural system: a strategy for rehabilitation of 'black soil type' deteriorated grassland on the Qinghai-Tibetan Plateau. In: Zheng D, Zhu L (eds) Formation and evolution, environment changes and sustainable development on the Tibetan Plateau. Academy Press, Beijing, pp 334–339 (in Chinese with English abstract)

Ma Y, Dong Q, Shi J (2001) Current situation and restoration of 'black soil type'degraded grassland in Qinghai Province. In: Sheng Z (ed) Qinghai livestock development project of technologies cooperation between EU and China. Qinghai Peoples Press, Xining, pp 90–99 (in Chinese)

Ma Y, Lang B, Li Q, Shi J (2002) Study on rehabilitating and rebuilding technologies for degenerated alpine meadow in the Changjing and Yellow River source region. Prataculture Sci 19(9):1–5 (in Chinese with English abstract)

Margules CR, Pressey RL (2000) Systematic conservation planning. Nat 405(6783):243

Miehe G, Miehe S, Kaiser K, Jianquan L, Zhao X (2008) Status and dynamics of the *Kobresia pygmaea* ecosystem on the Tibetan Plateau. Ambio 37(4):272–279

Miehe G, Bach K, Miehe S, Kluge J, Yongping Y, Duo L, Oc S, Wesche K (2011) Alpine steppe plant communities of the Tibetan highlands. Applied Vegetation Science doi: 10.1111/j.1654-109X.2011.01147.x

Millennium Ecosystem Assessment (2005) Ecosystems and human well-being: synthesis. Island Press, Washington

Pressey RL (1994) Ad Hoc reservations: forward or backward steps in developing representative reserve systems? Conserv Biol 8(3):662–668

Ren X (2008) Wild animals in the Sanjiangyuan can be seen anywhere. http://news.xinhuanet.com/mrdx/2008-10/06/content_10154148.htm. Accessed Nov 2008

Sheng G (2006) Studies on ecological immigrant of three river sources and sustained development. J Qinghai Natl Inst 32(1):109–112 (in Chinese with English abstract)

Shi X (2006) Analysis and strategies on advantage of developing organic animal husbandry in Qinghai. Qinghai Sci Technol 4:7–9 (in Chinese with English abstract)

Shi H, Wang Q, Jing Z (2005) The structure, biodiversity and stability of artificial grassland plant communities in the source regions of the Yangtze and yellow river. Acta Pratacultural Sci 14(3):23–30 (in Chinese with English abstract)

Smardon RC, Faust BB (2006) Introduction: international policy in the biosphere reserves of Mexico's Yucatan Peninsula. Landsc Urban Planning 74(3–4):160–192

Smith AT, Foggin JM (1999) The plateau pika (*Ochotona curzoniae*) is a keystone species for biodiversity on the Tibetan plateau. Anim Conserv 2:235–240

Soulé ME (ed) (1987) Viable populations for conservation. Cambridge University Press, Cambridge

Squires V, Hua L, Zhang D and Li G (eds) (2010) Towards sustainable use of rangelands in North-West China. Springer

Statistical Bureau of Qinghai (ed) (2003) Qinghai Statistical Yearbook. China Statistical Press, Beijing (in Chinese)

STRA-REP (2006) Revised terms of reference of the committee of experts for the development of the Pan-European ecological network. UNEP, Strasbourg (rep 06_02e rev/MB/bs). Sourced from: http://www.peblds.org/files/meetings/strarep_2006_04_en.pdf. Accessed Nov 2008

Sun H, Zheng D (eds) (1996) Formation and evolution of Qinghai-Tibetan Plateau. Shanghai Science and Technology Press, Shanghai (in Chinese with English abstract)

Tang X (2003) Basic ecological characteristics of the three-rivers source area and design of the nature reserve. For Res Manag 1:38–44 (in Chinese with English abstract)

The Nature Conservancy (2000) Osage Hills/Flint Hills ecoregional conservation plan. The nature conservancy, Topeka. Sourced from: www.conserveonline.org/library/final_plan_appendices.pdf. Accessed Nov 2007

Wang Q (2007) Climate resource features in recent 40 years in Sanjiangyuan region. In: Chen G-C, Chen X-Q, Gou X-J (eds) Ecological environment in Sanjiangyuan natural reserve. Qinghai People's Press, Xining, pp 43–85 (in Chinese)

Wang G, Cheng G (2001) Characteristics of grassland and ecological changes of vegetations in the source regions of Yangtze and yellow rivers. J Desert Res 21(2):101–107 (in Chinese with English abstract)

Wang C, Wang W (2004) Sustainable develop of population, resources, environment in three river fountainhead–concurrently discuss development of Qinghai Province nation area. Northwest Popul J 2:32–36 (in Chinese with English abstract)

Wang G, Shen Y, Cheng G (2000) Eco-environmental changes and causal analysis in the source regions of the yellow river. J Glaciol Geocryol 22(3):200–205 (in Chinese with English abstract)

Wang G, Qi L, Cheng G (2001) Climatic changes and its impact on the eco-environment in the source region of the Yangtze and yellow rivers in recent 40 years. J Glaciol Geocryol 23(4):346–351 (in Chinese with English abstract)

Wang Q, Shi H, Jing Z (2004) Recovery and benefit analysis of ecology on degraded natural grassland of the source region of Yangtze and yellow rivers. Pratacultural Sci 21(12):37–41 (in Chinese with English abstract)

Wang Q, Lai D, Jing Z, Li S (2005) The resources, ecological environment and sustainable development in the source regions of the Yangtze, yellow and Yalu-Tsangpo rivers. J Lanzhou Univ (Nat Sci) 41(4):31–37 (in Chinese with English abstract)

Wang X, Liu F, Qiang Z (2007) Integration issues studying with ecological immigration of Sanjiangyuan area. Ecol Econ 2:403–406 (in Chinese with English abstract)

Winkler D (2009) Caterpillar fungus (*Ophiocordyceps sinensis*) production and sustainability on the Tibetan Plateau and in the Himalayas. Asian Medicine 5(2):291–316

Xiao Z (2006) The government investment to continually strengthen ecological construction in Qinghai in the 11th five year plan. Pratacultural Sci 23(1):51 (in Chinese)

Xin H (2008) A green fervor sweeps the Qinghai-Tibetan Plateau. Science 321(5889):633–635

Zhang G, Li X, Li L, Hu L (1998) Meteorological analysis on the forming of barelands in cold highland grassland in Southern Qinghai Plateau. Grassland China 6:13–17,25 (in Chinese with English abstract)

Zhang G, Li L, Wang Q (1999) Effects of climatic changes of south Qinghai Plateau on the alpine meadow. Acta Pratacultural Sci 3:1–10 (in Chinese with English abstract)

Zhang T, Zhang Q, Zhang Z (2005) The choice of programme of ecological immigration and its follow up industry. Popul Sci China 1:28–33 (in Chinese with English abstract)

Zheng D (1996) Study on the natural section system of Qinghai-Tibetan Plateau. Sci China 26(4):336–341 (in Chinese with English abstract)

Zhou G, Yie G (2007) Wetland protection conditions and its sustainable development in Sanjiangyuan region. Inn Mong For Investig Des 30(5):14–15,57 (in Chinese with English abstract)

Zhou H, Zhou L, Liu W, Zhao X (2003a) The study on the reason of grassland degradation and the strategy of sustainable development of animal husbandry in Guoluo prefecture, Qinghai Province. Prataculture Sci 20(10):19–25 (in Chinese with English abstract)

Zhou H, Zhou L, Zhao X (2003b) The degraded process and integrated treatment of 'black soil beach' type degraded grassland in the source regions of Yangtze and yellow rivers. Chin J Ecol 22(5):51–55 (in Chinese with English abstract)

Zhou H, Zhao X, Tang Y, Gu S, Zhou L (2005) Alpine grassland degradation and its control in the source region of the Yangtze and Yellow rivers, China. Jpn Soc Grassland Sci 51(3):191–203

Zhu X, Dai S (2008) Reaching to 65% of vegetation coverage rate and increasing of around 1000 lakes in the yellow river source in Madou county. http://www.guoluo.gov.cn/html/116/6821.html. Accessed Oct 2008 (in Chinese)

Index

D. Higgitt (ed.), *Perspectives on Environmental Management and Technology in Asian River Basins*, SpringerBriefs in Geography, DOI: 10.1007/978-94-007-2330-6, © The Author(s) 2012